OFDMA for Broadband Wireless Access

For a listing of recent titles in the *Artech House Mobile Communications Series*, turn to the back of this book.

OFDMA for Broadband Wireless Access

Sławomir Pietrzyk

ARTECH HOUSE
BOSTON | LONDON
artechhouse.com

Library of Congress Cataloging-in-Publication Data
A catalog record of this book is available from the U.S. Library of Congress.

British Library Cataloguing in Publication Data
A catalog record of this book is available from the British Library.

ISBN: 1-59693-044-6

Cover design by Yekaterina Ratner

685 Canton Street
Norwood, MA 02062

International Standard Book Number: 1-59693-044-6

10 9 8 7 6 5 4 3 2 1

Contents

Preface

The requirements imposed on future broadband wireless access (BWA) systems include support of high data rates, operation in a hostile multipath radio environment, provision of various QoS profiles, and minimal consumption of the limited radio resources, such as RF bandwidth and transmit power. One of the most promising multiple access technologies that satisfies these demands is orthogonal frequency division multiple access (OFDMA). This technology has recently gained significant interest due to the standardization work resulting for example in specifications of the IEEE 802.16 standard (also known as WiMAX). In this book, the problems of radio resource management for OFDMA systems are addressed, with particular interest on three OFDMA system cases, namely: (1) with no spectrum (subcarriers) reuse, (2) with intracell spectrum reuse by applying a so-called embedded modulation and (3) with intercell (cellular) spectrum reuse. The two latter system cases are bound by the common presence of cochannel interference caused by reuse of subcarriers.

The main goal of this book is to present and analyze recently developed radio resource management algorithms for the downlink of OFDMA multiple access systems, which include algorithms for: base station assignment, subcarrier allocation, bit loading, and transmit power control. The performance of the described schemes is further evaluated in numerical experiments settled in the WiMAX scenario.

Chapter 1 of this book introduces the reader into the area of BWA systems. After presentation of the requirements imposed on such systems, a historical perspective of BWA development is outlined, which is followed by brief characterization of the wireless systems employing OFDM/OFDMA. Also, a more detailed book preview is included in this chapter.

Chapter 2 introduces the characteristics of the radio channel and describes its influence on the design of radio communication systems. As one of the most efficient schemes to handle multipath propagation caused by the radio medium, OFDM is described. This is followed by introducing OFDMA that is a multiple access scheme based on OFDM.

In Chapter 3 we describe the principles of radio resource management in wireless access systems, addressing issues such as multiple access, QoS, and resource reuse. Then we outline the approach towards radio resource management, which is taken in this book.

In Chapter 4 we provide an overview of the IEEE 802.16 standard. Also in this chapter we explain how the IEEE 802.16 scenario is used in the following chapters.

Chapter 5 is concerned with radio resource management algorithms for single-cell downlink OFDMA systems, where subcarriers are exclusively allocated to users (one subcarrier can be at most allocated to one user). Exclusive allocation simplifies the resource assignment process, since cochannel interference is not present.

In Chapter 6, single-cell downlink OFDMA system is considered, where subcarriers are reused within the same cell by the application of embedded modulation, known so far mainly from broadcast systems. It allows to stack multiple signals that can be addressed to different users on a single subcarrier, creating in this way an increased number of independent links and thus increasing the system's capacity. Stacking multiple signals, which increases embedded modulation order, may also be perceived as a generalization of adaptive modulation, where bits loaded on a single subcarrier are not necessarily addressed to one user but may be addressed to various users, depending on the underlying channel conditions. This extra freedom in resource allocation makes embedded modulation a more power-efficient scheme than conventional adaptive modulation.

Chapter 7 is concerned with intercell reuse of subcarriers in a multiple-cell (cellular) downlink OFDMA system. Cellular system is a basic prerequisite to have OFDMA applied widely as a technique for next generation networks. In such a system case, a reuse factor of one is applied (i.e., the entire bandwidth is available in every cell and the selection of a particular subcarrier is subject to local load and channel conditions). The combination of adaptive subcarrier assignment, adaptive modulation, and power control handling CCI leads to substantial gains in total transmit power reduction. Provision and maintenance of the required QoS are the key assumptions taken in the development of these methods.

In Chapter 8, we conclude with a summary of presented solutions and results. We indicate potential applications for OFDMA as a future multiple access scheme for BWA systems. Also, we review other recently considered technologies improving BWA systems performance such as multiple antennas or multiuser detection.

Finally, we discuss the main future research and standardization directions and cast some long-term views on the future design of BWA systems.

This book is based on my Ph.D. thesis defended in the Wireless and Mobile Communications Group at the Delft University of Technology in the Netherlands. I would like to thank all the reviewers of my Ph.D. thesis and members of my doctoral committee: Prof. J.C. Arnbak, Prof. H. Steendam, Prof. E.R. Fledderus, Prof. W.C. van Etten and in particular my promotors, Dr. G.J.M. Janssen and Prof. I.G.M.M. Niemegeers. I would like to thank all the members of the Ubiquitous Communications Project at the Delft University of Technology, where this work was initiated. I would also like to thank the entire Artech House team, and in particular Judi Stone, Kath Nolan and Tiina Ruonamaa, for guiding me through the process of book development. Last but not least I thank my wife Kinga, our daughter Agata and our sons Maciek and Bartek for their love and patience.

Figure 4.1 and Table 4.1 reprinted with permission from IEEE Std 802.16-2004 "Part 16: Air Interface for Fixed Broadband Wireless Access Systems" Copyright ©2004, by IEEE. The IEEE disclaims any responsibility or liability resulting from the placement and use in the described manner. IEEE, 802.16, WirelessMAN-OFDM, WirelessMAN-OFDMA, WirelessMAN-SC, and WirelessMAN-SCa, are trademarks owned by the Institute of Electrical and Electronic Engineers Inc. OFDMA is a registered trademark of Runcom Technologies Ltd. The author would like to acknowledge Adrian Bohdanowicz for simulation software used for channel modeling and Dr. Christophe Fiorio (http://www.lirmm.fr/~fiorio/) for the algorithm2e.sty package used in this book for algorithms compilation.

Chapter 1

Introduction to broadband wireless access

What is invisible, often attracts people's attention. When I was a child my father showed me how to build a small toy that I could use to listen to radio programs out of the ... central heating system. Though it may sound bizarre, this is exactly how I perceived it at the time. As appeared later, it was a simple AM radio receiver based on a diode detector and an earphone that, when connected to the heating radiator, served as an antenna for long waves together with the entire heating installation. Hearing weak voices of the radio programs known from the regular radio receiver was truly amazing. This invisible presence of voices still remains in my childhood memories.

Nowadays, radio is part of numerous communication systems, including broadcast, transport, and access. The access systems, being of particular interest in this book, are stretching from short-range personal-size networks (such as Bluetooth) to long-range metropolitan-size networks (such as IEEE 802.16 specification [1]). What was yesterday's miracle becomes today's or tomorrow's reality. Innovations and changes in different areas of technology and economy make it possible to build systems that can help people to satisfy their basic needs of interpersonal communication and access to information, while at the same time enjoying the freedom of wireless operation. These innovations include the shift towards digitalization of information, advances in microelectronics, decreasing cost of digital signal processing, convergence of the computer and telecommunications industries, deregulation of the telecommunication markets in many countries, and the extensive standardization efforts. All these changes contribute to the benefits of the economies of scale and lead to a situation that may seem contradictory at first glance but is real at the second: communication and computing devices have become smaller, cheaper AND more powerful [2]. Because of this, what was not so long ago a bulky and

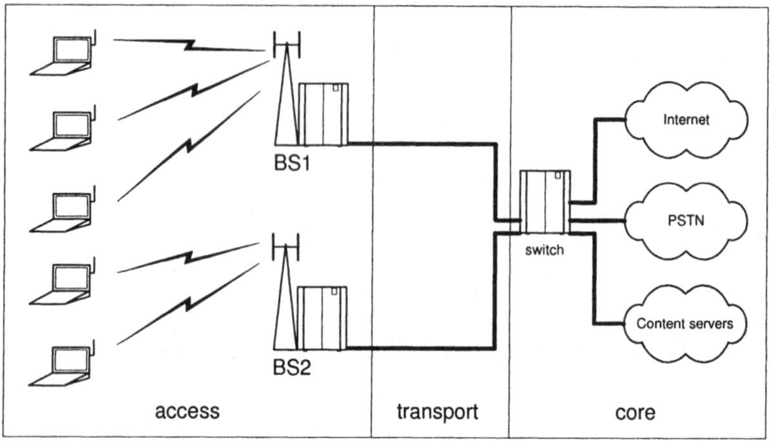

Figure 1.1 Access, transport, and core parts of a communication network.

expensive piece of hardware (a computer or a communicator) can now be carried in one's bag or even a pocket, at a reasonable price [3]. Personalization goes well with wireless because it would be difficult to imagine handy personal devices, each with at least one cable connected. Therefore, wireless is particularly suitable for systems directly serving the end-customer, such as access systems. An example access system is depicted in Figure 1.1. End-customer (subscriber) station is connected via a wireless link to the base station (BS), which belongs to the access network. The access network is connected to the core network using the transport network. Although in principle, wireless technologies can be used in all network subsystems, most often they are employed in the access subsystem. By providing a cost-efficient and easy-to-deploy infrastructure for connecting end-customers, which is usually not the case for cable systems such as copper, coaxial, or fiber, wireless access systems allow to solve the so-called last mile problem. This is especially visible in rural or developing areas. Nothing comes without a price and wireless systems are probably most challenging to design, which is mainly due to the difficult nature of the radio medium. Moreover, the proximity of the end-customer and what follows, the multitude of different expectations and requirements, calls for extreme flexibility of the wireless access system. And of course, we want our systems to be *broadband*. Although this term is very much overused, it carriers at least two widely recognized features: high speed and universal. These simply translate into a comfortable experience of the end-customer when using communication services and the ability to tailor multiple services in order to better respond to customers' expectations and be fast-to-market. The quantification of high speed, that is high data rates, is difficult to be determined, since what is high speed today is unlikely to be high speed tomorrow. By analogy, a similar situation happens to processors' speeds for personal computers. Universality of broadband wireless access systems can be understood as their ability to support various services such as Internet access, e-mail, digital television, or voice communication. Thanks to broadband, a new service may be formed without the need to build a new network.

In the following section, we outline the key requirements that are imposed on future broadband wireless access (BWA) systems.

1.1 REQUIREMENTS IMPOSED ON BWA SYSTEMS

As with every technological innovation, modern broadband wireless access systems cannot be designed without considering characteristics of the environment in which they need to operate and the requirements imposed on the services these systems should provide. Below, we briefly list some of those desired key features:

- High data rates and support of different traffic profiles, also referred to as quality-of-service (QoS) profiles [4],
- High capacity (i.e., the ability to serve a high number of users),
- Efficient use of available resources, which means satisfying the increased requirements by making better use of the already available resources (e.g., spectra) than by allocating new ones,
- Operation in a harsh and changeable environment of the radio channel, in which variability should be treated as usual rather than exceptional [5],
- Low power consumption (which applies to the construction of algorithms and the design of processors) and low power emission (meaning low levels of transmit power),
- Freedom of personalization and mobility. In [5] the notion of *nomadicity* (nomadic computing and communications) is addressed, which is basically a provision of communication services to the nomad (i.e., a person on the move),
- (Co)operation of a number of different communication devices as well as entire systems at the same time and at the same place [4]. This requires

the ability to handle the mutual influence of devices or systems, which undoubtedly happens in such situations either by intentional or unintentional interactions. This influence is most often referred to as interference (usually cochannel interference).

One of the most promising multiple access techniques, which satisfies the demands presented above, is orthogonal frequency division multiple access (OFDMA) [6–8]. OFDMA is based on orthogonal frequency division multiplexing (OFDM) [9, 10], a transmission scheme that partitions the available bandwidth into N narrowband subcarriers. In order to overcome frequency selective fading causing intersymbol interference, the bandwidth of each subcarrier is chosen to be sufficiently smaller than the coherence bandwidth of the channel. Therefore, OFDM is extremely robust to multipath propagation, which is one of the phenomena causing significant problems in wireless communications. In addition, the narrowband OFDM subcarriers are placed very close to each other, which results in a high spectral efficiency. OFDMA is in fact a form of frequency division multiple access (FDMA), where a user may be assigned one or more subcarriers (equivalent to FDMA frequency channels) in order to satisfy its traffic requirements. The key advantage of OFDMA is that it allows for *multiuser diversity*: a subcarrier that is of low quality to one user can be of high quality to another user and can be allocated accordingly [11]. In this way, a subcarrier is left unused only if it is of low quality to all users. Moreover, this approach can be combined with adaptive modulation, where modulation levels are chosen on a per-subcarrier basis according to the observed channel conditions in order to further increase spectral efficiency. Therefore OFDM/OFDMA is a key element of the majority of the modern communication systems, including BWA, cable access (xDSL), broadcasting (DVB, DAB), and WLANs (802.11, HiperLAN).

In the following sections, after presenting a historical perspective of the BWA systems' development, we shall concentrate on a family of IEEE 802.16 standards [1, 12, 13] commonly referred to as the Worldwide Interoperability for Microwave Access (WiMAX) system. In this book, we will use the term WiMAX 802.16 interchangeably. In the next section, other communication systems applying OFDM/OFDMA will be reviewed, such as WLAN or broadcast systems.

1.2 HISTORICAL PERSPECTIVE OF BWA SYSTEMS DEVELOPMENT

In the early days of BWA systems development, the only truly broadband service was television. Also here, the last-mile problem existed and engineers tried to solve it by reaching end-customers by use of radio rather than cable. In the following, we shall introduce two early BWA systems, namely multipoint multichannel distribution system (MMDS) and local multipoint distribution system (LMDS). Because this book is primarily focused on fixed wireless access, we do not address mobile systems such as UMTS or HSDPA [14].

MMDS was historically one of the first BWA systems [15] present on the market. It offers maximally 33 analog video channels within a bandwidth of 500 MHz [16]. It operates in frequencies around 2.5 GHz and can be treated as a wireless realization of cable television (CATV). MMDS systems in their evolved versions based on 64-QAM can also provide the "last mile" for wireless access based, for example on ATM.

LMDS, which is a digital successor of MMDS, is designed to operate in 27-40 GHz frequency bands and can provide high speed wireless data access [17]. Operation in millimeter frequencies requires the presence of line-of-sight (LOS) between the base station and user equipment. Multipath is reduced by using narrowbeam sectorized antennas. The LOS requirement often makes the customer antenna installation difficult, which limits universality of this system. LMDS offers maximal cell ranges of about 1-5 km, with uplink data rates up to 8 Mbit/s and downlink rates (shared) around 36 Mbit/s [17, 18].

The advantages of the early BWA systems based on MMDS or LMDS include mainly a relatively low cost and fast deployment, when compared to their cable counterparts. However, both MMDS and LMDS require LOS and do not provide sufficient indoor coverage due to weak in-building penetration. This practically limits the system's operation to applications such as backhauling, leaving individual household access for systems that do not require LOS. Shift from LOS operation to non-LOS (NLOS) operation is visible in 802.16 standardization development [1], where first releases [12] bear strong resemblance to LMDS. Chapter 4 discusses the 802.16 family of standards in detail.

It is worth mentioning that BWA systems (terrestrial) are competing with such access technologies as ADSL, VDSL, satellite and fiber, as well as mobile systems including CDMA, UMTS, and HSDPA [19].

1.3 OVERVIEW OF SYSTEMS EMPLOYING OFDM/OFDMA

Thanks to its major merits of robustness to multipath propagation, bandwidth efficiency, and flexibility in partitioning the spectral resource, OFDM is a technique applied in various contemporary communication systems, including digital audio

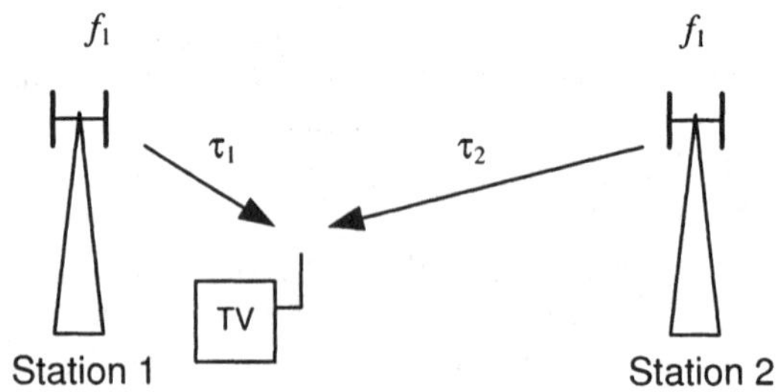

Figure 1.2 The concept of single-frequency network applied to TV broadcasting. The two stations are broadcasting using the same frequency. TV receiver receives two delayed (τ_1 and τ_2) versions of the broadcasted signal. The delay is resolved by the application of OFDM.

and video broadcasting (DAB and DVB) [20,21], wireless LANs (WLAN) [22–26], and BWA systems such as [12,27,28], or [29]. OFDM, known also as a multicarrier modulation (MCM), is applied in cable access networks such as asymmetrical digital subscriber lines (ADSL), hybrid digital subscriber lines (HDSL) [30], and was considered for cable television access [8]. A good review of DSL systems is presented in [31] and a thorough discussion on the application of OFDM in ADSL is given in [32].

In the following, a review of contemporary wireless systems implementing OFDM/OFDMA is given, with a focus on the issues of multiple access and resource management.

1.3.1 Broadcast systems

The first standard applying OFDM was the one for digital audio broadcasting (DAB) [20]. DAB offers high quality audio (comparable to CD quality) and is a successor of the current analog broadcast systems based on AM or FM. The role of OFDM in DAB is twofold. First, it helps to reduce the effect of multipath, which is present in a usually NLOS broadcast environment. Second, it realizes a single frequency network, where every transmitting station uses the same frequency band, which allows for efficient use of spectrum and simple network planning. This is schematically depicted in Figure 1.2. A receiver may be in the range of multiple stations, which provides gain due to macro-transmit diversity. The propagation delays from different stations are treated as multipath components of the same original signal and are resolved by the application of a cyclic prefix. DAB is standardized in four different modes, which differ in the separation allowed between the transmitters (from 12 to 96 km). This separation corresponds to different delay spreads, which are taken into account in determination of such OFDM system parameters as the number of subcarriers and the duration of a symbol and a cyclic prefix.

Also in case of the digital video broadcasting - terrestrial (DVB-T) standard [21], robustness to multipath and realization of a single frequency network are key motives that decided in favor for using OFDM in the PHY layer. DVB-T is standardized in two modes, namely a 2k and an 8k mode, where the name corresponds to the number of subcarriers. In the 2k mode, the system operates 1705 subcarriers, while in the 8k mode the number of subcarriers is 6817. The following modulation schemes are allowed: QPSK, 16-QAM, 64-QAM and, MR-QAM (multiresolution QAM). The last scheme is closely related to the embedded modulation as considered in Chapter 6 of this book.

1.3.2 Wireless LAN systems

The next group of standardized wireless systems implementing OFDM are wireless LAN solutions. There are two major standards at present that are very similar to each other regarding the physical (PHY) layer, namely IEEE 802.11a [22] (with its relative IEEE 802.11g [23]) and ETSI HiperLAN/2 [24, 25]. Both operate in the 5-GHz unlicensed national information infrastructure (UNII) band, and both use 64 subcarriers and apply adaptive modulations such as BPSK, QPSK, 16-QAM and, 64-QAM. Major differences are in the medium access control (MAC) layer, where in the case of 802.11a the Ethernet-like "listen-before-talk" approach of CSMA/CA is applied, while in the case of HiperLAN/2 a connection-oriented TDMA/TDD based on wireless ATM is used. Moreover, HiperLAN/2 has implemented radio resource management features such as:

- Dynamic frequency selection, which allows to choose the best suitable channel for transmission based on measurements and to avoid channels that are interfered,
- Power control, which reduces interference to other systems.

These methods are obligatory by European regulations and they are added to the 802.11 standard in 802.11h.

1.3.3 BWA systems

Recently, BWA systems have gained considerable interest. This interest is probably most visible in the IEEE 802.16 standardization and therefore the IEEE 802.16 is treated in a separate chapter of this book (Chapter 4). However, there are also some other activities of various companies, which, in effect, are also related to the WiMAX family.

One of them is Flarion, which developed a BWA system employing OFDM [27]. The applied solution known as Flash-OFDM is a combination of OFDM with fast-frequency hopping performed over OFDM subcarriers. Since the Flash-OFDM system is neither open nor standard, more detailed information is not available.

Another example is software defined OFDMA/TDD broadband wireless access platform offered by Adaptix Inc. [29]. The system is based on 802.16e and WiBro, which is a broadband wireless access system developed in Korea.

1.3.4 Summary of the key parameters of the contemporary OFDM(A) systems

A summary of the basic parameters of the contemporary OFDM/OFDMA systems discussed is given in Table 1.1. It can be concluded that OFDM/OFDMA are used in a broad variety of systems, which results in different values of key system parameters, such as the number of OFDM subcarriers or frequency bands in which a system operates.

1.4 PREVIEW OF THE BOOK

The WiMAX system is of great interest now to standardization bodies, infrastructure manufacturers, operators, and research institutions. One of its most promising PHY layers is based on OFDMA. Algorithmic approach to OFDMA (as well as to any modern wireless access systems) is gaining momentum. This is because most of the intelligence can now be put into a DSP and the system performance (expressed e.g., in offered data rates, number of served users, or power consumption) so important for the operator and for the end customer depends directly on how efficiently the available resources are managed by these RRM algorithms. On the other hand, radio resource management mechanisms implemented in contemporary telecommunications systems utilizing OFDM/OFDMA are not fully exploiting its great potentials. Also, scientifically, the area of radio resource management for OFDMA is relatively

Table 1.1

Summary of Basic Parameters of Contemporary Systems Utilizing OFDM(A)

System	Number of subcarriers N	System Bandwidth BW	Modulation schemes	Carrier Frequency f_c
DAB	192, 384, 768, 1536	1.5 MHz	Differential QPSK	$\sim 0.2 - 3$ GHz
DVB-T	1705, 6817	8 MHz	QPSK, 16-QAM, 64-QAM, MR-QAM	$\sim 470 - 682$ MHz
802.11a	64	20 MHz	BPSK, QPSK, 16-QAM, 64-QAM	~ 5 GHz
HiperLAN/2	64	20 MHz	BPSK, QPSK, 16-QAM, 64-QAM	~ 5 GHz
WirelessMAN-OFDM	256	$\sim 1.5 - 28$ MHz for licensed bands, $\sim 10 - 20$ MHz for unlicensed bands	BPSK, QPSK, 16-QAM, 64-QAM	$\sim 2 - 11$ GHz
WirelessMAN-OFDMA	2048	$\sim 1.5 - 28$ MHz for licensed bands, $\sim 10 - 20$ MHz for unlicensed bands	QPSK, 16-QAM, 64-QAM	$\sim 2 - 11$ GHz
Flash-OFDM	?	2x 1.25 MHz	QPSK, 16-QAM, 64-QAM	$\sim 0.4 - 3$ GHz

unexplored with many open issues. The basic motivation of this book is to present state-of-the-art radio resource allocation algorithms for OFDMA, which satisfy the requirements of operation in a harsh radio environment, support and maintain various (and demanding) QoS profiles, and which exploit the following features:

- Multiuser diversity and adaptive subcarrier allocation: a subcarrier that is of low quality to one user may be of high quality to another user and should be allocated accordingly,
- Per-subcarrier (or per block of subcarriers) adaptation of modulation modes and power control,

- Intracell subcarrier reuse: a subcarrier may be of high quality for more than one user thus it may be power efficient to allocate this subcarrier to more than one user within the same cell,
- Intercell subcarrier reuse in a cellular system with cooperative power control: the transmit power level for a particular subcarrier is determined together for all cochannel users; in this way an equilibrium among the cochannel users is achieved.

This book addresses the problems of RRM and QoS for OFDMA systems and discusses possible solutions presented in a form of RRM algorithms. In the case of OFDMA systems, RRM boils down to the determination of methods for allocation of subcarriers to users, power control, and so-called bit loading (setting the number of bits per each subcarrier). Additionally, in a cellular environment, assignment of users to base stations should be performed. Of particular interest are scenarios where interference caused by spectrum reuse is present. Such scenarios where available spectrum is reused are unavoidable if we think about a widespread use of OFDMA as a technique for the next generation networks.

1.4.1 Book outline

The book is organized as follows. Chapter 2 gives more insight into the characteristics of the multipath radio channel and provides a theoretical background for OFDM (treated as a transmission technique), which forms the basis for the PHY layer operation of OFDMA.

Chapter 3 explains principles of radio resource management and QoS support. It also summarizes main features of contemporary BWA systems and introduces the resource allocation tasks described further in this book.

Chapter 4 summarizes the key points of the IEEE 802.16 standard, concentrating on characterization of the PHY and MAC layers. Also in this chapter we introduce a system scenario and key system parameters based on 802.16, which are used later in this book in the system's analysis and simulations.

Chapter 5 considers a single-cell OFDMA system with exclusive allocation of subcarriers to users (no spectrum reuse) and without cochannel interference, which is the simplest system case described in this book. This chapter presents a broad review and classification of the most useful resource allocation algorithms that can be directly applied for OFDMA with exclusive subcarrier allocation, and that can also form the basis for designing more advanced algorithms applicable to situation where cochannel interference is present. This chapter also introduces the traffic model that is used in numerical experiments further in this book.

The problem of radio resource allocation for OFDMA systems reusing subcarriers within a single cell (intracell spectrum reuse) by the application of embedded modulation is studied in Chapter 6. Efficient algorithms for resource allocation for OFDMA systems utilizing embedded modulation are described and evaluated.

The resource allocation problem for cellular OFDMA systems with intercell reuse of subcarriers is considered in Chapter 7. In this chapter, the RRM algorithms for base station assignment, subcarrier allocation, bit loading, and power control for the downlink of cellular OFDMA system are presented together with their thorough performance evaluation. Radio resources are distributed efficiently according to user individual QoS requirements and the QoS is maintained during the entire lifespan of all involved communication links. Here, the available bandwidth is aggressively reused in the system, where in principle the entire bandwidth is available at every cell and the selection of a particular subcarrier is subject to local load conditions. In this way, a network that is self-organizing in frequency reuse is realized.

Finally, in Chapter 8 we summarize main conclusions of the previous chapters, provide recommendations for future research, and cast a view on the future of OFDMA. In addition, other new technologies are described that may play a role in future improvements of BWA standards.

References

[1] IEEE, *Part 16: Air Interface for Fixed Broadband Wireless Access Systems,* IEEE Std 802.16-2004, October 2004.

[2] Gatherer, et al., "DSP-Based Architectures for Mobile Communicaions: Past, Present and Future," *IEEE Communications Magazine,* January 2000, pp. 84-90.

[3] Viterbi, A.J., "Wireless Digital Communications: A View Based on Three Lessons Learned," *IEEE Communications Magazine,* September 1991, pp. 33-36.

[4] Ohmori, S. , Yamao, Y., and Nakajima, N., "The Future Generations of Mobile Communications Based on Broadband Access Technologies," *IEEE Communications Magazine,* December 2000, pp. 134-142.

[5] Kleinrock, L., "On Some Principles of Nomadic Computing and Multi-Access Communications," *IEEE Communications Magazine,* July 2000, pp. 46-50.

[6] Wei, L., *Single-user and Multi-user Modulation and Detection Methods for Mobile Communications,* University of South Australia, PhD Dissertation, 1994.

[7] Wei, L., and Schlegel, C., "Synchronization Requirements for Multi-user OFDM on Satellite Mobile and Two-path Rayleigh Fading Channels," *IEEE Transactions on Communications,* Feb/Mar/Apr 1995, volume 43, pp. 887-895.

[8] Sari, H., Levy, Y., and Karam, G., "Orthogonal Frequency-Division Multiple Access for the Return Channel in CATV Networks," Proc. ICT, 1996.

[9] *Orthogonal frequency division multiplexing,* U.S. Patent 3,488,445, filed Nov. 14, 1966, issued Jan. 6, 1970.

[10] Weinstein, S.B., and Ebert, P.M., "Data Transmission by Frequency-Division Multiplexing Using the Discrete Fourier Transform," *IEEE Transactions on Communications Tech.,* October 1971, volume 19, pp. 628-634.

[11] Wong, C.Y., et al., "Multiuser OFDM with Adaptive Subcarrier, Bit, and Power Allocation," *IEEE Journal on Selected Areas in Communications,* October 1999, volume 17, pp. 1747-1758.

[12] IEEE, *Part 16: Air Interface for Fixed Broadband Wireless Access Systems,* IEEE Std 802.16, 2001.

[13] IEEE, *Part 16: Air Interface for Fixed Broadband Wireless Access Systems-Amendment 2: Medium Access Control Modifications and Additional Physical Layer Specifications for 2-11 GHz,* IEEE Std 802.16a (Amendment to IEEE Std 802.16-2001), 2003.

[14] Holma, H., and Toskala, A., (eds.), *WCDMA for UMTS : Radio Access for Third Generation Mobile Communications,* Wiley, 2004.

[15] Gorshe, S.S., and Papir, Z., "Non-Line-of-Sight Broadband Wireless Access Systems," guest editorial, *IEEE Communications Magazine,* April 2002, pp. 84-85.

[16] Honcharenko, W., et al. "Broadband Wireless Access," *IEEE Communications Magazine,* January 1997, pp. 20-26.

[17] Nordbotten, A., "LMDS Systems and Their Applications," *IEEE Communications Magazine,* June 2000, pp. 150-154.

[18] Mahonen, P., et al., "Wireless Internet over LMDS: Architecture and Experimental Implementation," *IEEE Communications Magazine,* May 2001, pp. 126-132.

[19] Drewes, C., Aicher, W., and Hausner, J., "The Wireless Art and the Wired Force of Subscriber Access," *IEEE Communications Magazine,* May 2001, pp. 118-124.

[20] ETSI, *Radio broadcasting systems: Digital Audio Broadcasting (DAB) to mobile, portable and fixed receivers,* ETS 300 401 ed.2, May 1997.

[21] ETSI, *Digital video broadcasting: Framing structure, channel coding and modulation for digital terrestrial television,* European Telecommunications Standard, EN 300-744, August 1997.

[22] IEEE, *Part 11: Wireless LAN Medium Access Control (MAC) and Physical Layer (PHY) Specifications. High-speed Physical Layer in the 5 GHz Band,* IEEE Std 802.11a, 1999.

[23] IEEE, *Part 11: Wireless LAN Medium Access Control (MAC) and Physical Layer (PHY) Specifications. Amendment 4: Further Higher Data Rate Extension in the 2.4 GHz Band,* IEEE Std 802.11g, 2003.

[24] ETSI, *ETSI TS 101 475. Broadband Radio Access Networks (BRAN); HIPERLAN Type 2; Physical (PHY) layer,* December 2001.

[25] ETSI, *ETSI TS 101 761-2. Broadband Radio Access Networks (BRAN); HIPERLAN Type 2; Data Link Control (DLC) Layer; Part 2: Radio Link Control (RLC) sublayer,* January 2002.

[26] Johnsson, M., "HiperLAN/2 - The Broadband Radio Transmission Technology Operating in the 5 GHz Frequency Band," HiperLAN/2 Global Forum, 1999.

[27] http://www.flarion.com, Flash-OFDM.

[28] http://ieee802.org/20, IEEE 802.20.

[29] http://www.adaptix.com.

[30] Chow, P.S., Tu, J.C., and Cioffi, J.M., "A Discrete Multitone Transceiver System for HDSL Applications," *IEEE Journal on Selected Areas in Communications,* August 2001, volume 9, pp. 895-908.

[31] Starr, T., Cioffi, J.M., and Silverman, P., *Understanding Digital Subscriber Line Technology,* Prentice Hall, 1998.

[32] Saltzberg, B.R., "Comparison of Single-Carrier and Multitone Digital Modulations for ADSL Applications," *IEEE Communications Magazine,* November 1998, pp. 114-121.

Chapter 2

Radio channel characteristics and OFDM/OFDMA basics

Communications over the radio channel can be considered a challenging task. The main reason is that the propagating radio signal undergoes various physical phenomena, including reflection, diffraction, and scattering [1–3]. This is caused by the presence of objects between the transmitter and the receiver, such as buildings or forests in the outdoor environment, or walls and furniture in the indoor environment, which affect the propagation of radio waves and in general causes the signal to arrive at the receiver via multiple paths. In effect, the received signal may be significantly different than the transmitted one, which is especially visible in non-line-of-sight (NLOS) scenarios when compared to line-of-sight (LOS), Figure 2.1. Such a situation requires special precautions to be taken in order to decode correctly the transmitted information.

In this chapter, characteristics of the transmission medium (i.e., the multipath radio channel) are given, which are occasionally based on [4]. This is followed by the explanation of one of the most efficient techniques for transmission in a multipath radio channel, namely orthogonal frequency division multiplexing (OFDM) and of the related multiple access scheme, orthogonal frequency division multiple access (OFDMA). The presented analytical description is based on the work done at the Lulea University in Sweden [5, 6]. Finally, a discussion of major advantages and disadvantages of both OFDM and OFDMA is presented.

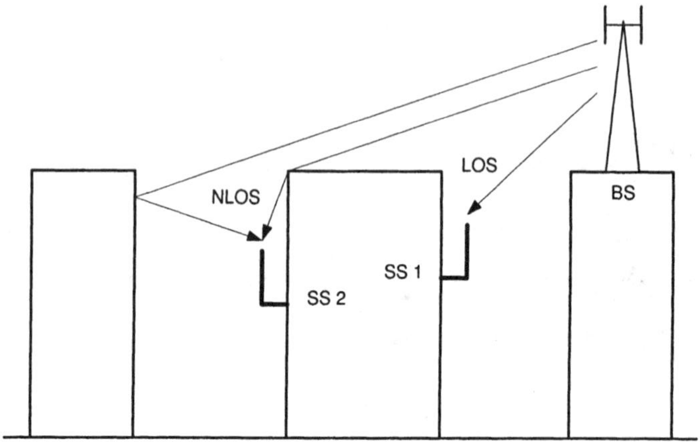

Figure 2.1 NLOS and LOS propagation in an urban wireless access system.

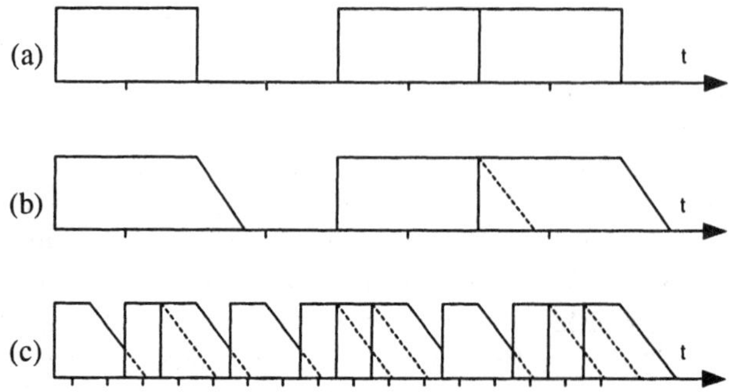

Figure 2.2 Influence of intersymbol interference (ISI): (a) low-speed data stream not affected by ISI, (b) low-speed data stream slightly affected by ISI, and (c) high-speed data stream significantly corrupted by ISI.

2.1 RADIO PROPAGATION

For carrier frequencies above 1 GHz, which are typical for a majority of the contemporary wireless access systems, reflections and shadowing are dominant phenomena [1, 2, 4]. In effect, the signal arriving at a receiver is a superposition of replicas of the transmitted signal, each with different amplitude, phase and delay time. This so-called multipath propagation results in dispersion of the received signal in the time domain and in frequency selectivity in the frequency domain. If sufficiently high data rates are used, the time dispersion (known as delay spread) causes transmitted data pulses to overlap, resulting in intersymbol interference (ISI), as depicted in Figure 2.2. In the next subsection, we characterize the multipath propagation. This is followed by a description of some other considered channel effects, namely path loss and shadowing.

2.1.1 Representation of the multipath channel

The multipath channel can be modeled as a linear filter described by the following complex equivalent baseband time-varying impulse response [7]

$$h(\tau; t) = \sum_{p=1}^{P} \beta_p(t) e^{j\Theta_p(t)} \delta(\tau - \tau_p(t)), \tag{2.1}$$

where: t is the observation time, τ is the delay, P is the number of multipath components, $\beta_p(t)$ is the amplitude, $\Theta_p(t)$ is the phase, and $\tau_p(t)$ is the delay of the path p. Assuming that the channel impulse response is time-invariant, it can then be expressed by the following relation

$$h(t) = \sum_{p=1}^{P} \beta_p e^{j\Theta_p} \delta(t - \tau_p). \tag{2.2}$$

The received equivalent low-pass signal is a sum of multiple replicas of the transmitted signal, which can be expressed by [8]

$$y(t) - x(t) \otimes h(t) = \sum_{p=1}^{P} \beta_p(t) e^{-j2\pi f_c \tau_p(t)} x(t - \tau_p(t)), \tag{2.3}$$

where

$$x(t) = Re\left\{ s(t) e^{j2\pi f_c t} \right\} \tag{2.4}$$

is the transmit signal with $s(t)$ being the equivalent low-pass transmit signal.

An often used measure for time dispersion of the channel is the rms (root mean square) delay spread, τ_{rms}. It is defined as the square root of the second central moment of the power delay profile [2]

$$\tau_{rms} = \sqrt{\int_{-\infty}^{\infty} \frac{(t - \tau_m^2)PDP(t)}{\overline{G}_c} dt}, \tag{2.5}$$

where τ_m is the average delay time defined by

$$\tau_m = \int_{-\infty}^{\infty} \frac{t \cdot PDP(t)}{\overline{G}_c} dt, \tag{2.6}$$

$\overline{G}_c$ is the average channel or power gain defined by

$$\overline{G}_c = \int_{-\infty}^{\infty} PDP(t)dt = \sum_{p=1}^{P} \beta_p^2, \tag{2.7}$$

and finally the power delay profile $PDP(t)$ can be defined as

$$PDP(t) = \sum_{p=1}^{P} \beta_p^2 \delta(t - \tau_p). \tag{2.8}$$

Time-dispersive channels can be often considered frequency-selective. This is because at the same time a signal could be faded at one frequency but not necessarily at another [9]. In order to model frequency selectivity, the channel response in the frequency domain $H(f)$ is calculated by taking the Fourier transform of the channel impulse response $h(t)$, which can be expressed by

$$H(f) = FT\{h(t)\} = \sum_{p=1}^{P} \beta_p e^{-j(2\pi f \tau_p - \Theta_p)}. \tag{2.9}$$

The measure for frequency selectivity is the channel *coherence bandwidth*, which can be expressed as

$$B_{coh} \approx \frac{1}{\xi \tau_{rms}}, \tag{2.10}$$

where ξ is a constant ($\xi = 2\pi$ in case of exponentially decaying PDP) [10]. The coherence bandwidth is the bandwidth over which different frequencies undergo the same attenuation. In other words this is the bandwidth over which the channel can be considered constant. If the bandwidth of the transmitted signal is larger than the coherence bandwidth then the channel is referred to as *frequency selective.* Otherwise, the channel is called *frequency nonselective* or *flat.*

The amplitude of the received multipath signal is often described by a Rician or a Rayleigh distribution [1]. When a significant LOS component is present then the channel can be modeled using a Rician distribution, with the PDF of the instantaneous received signal amplitude y_i given by [8,9]

$$f_{Rice}(y_i) = \frac{y_i}{\sigma_s^2} \exp\left(-\frac{y_i^2 + A^2}{2\sigma_s^2}\right) I_0\left(\frac{y_i A}{\sigma_s^2}\right), \tag{2.11}$$

where A is the amplitude of the dominant LOS component, σ_s^2 is the average power scattered over the reflected paths and $I_0(\cdot)$ indicates the zero-order Bessel function of the first kind and $y_i \geq 0$. The parameter describing the ratio of the dominant LOS path component versus the multipath components is known as Rician K-factor and can be expressed as [2]

$$K_{factor} = \frac{P_{dominant}}{P_{reflected}} = \frac{A^2}{2\sigma_s^2} = \frac{\beta_1^2}{\sum_{p=2}^{P} \beta_p^2}, \tag{2.12}$$

where $P_{dominant}$ is the power of the dominant path $p = 1$, and $P_{reflected}$ is the total scattered power.

When no LOS component (NLOS) is present (i.e., $A = 0$), the K_{factor} equals to zero and the Rician distribution simplifies to the Rayleigh distribution. The PDF of the instantaneously received power for Rayleigh fading channel is given then by [1,8]

$$f_{Rayleigh}(y_i) = \frac{y_i}{\sigma_s^2} \exp\left(-\frac{y_i^2}{2\sigma_s^2}\right). \tag{2.13}$$

Example PDFs of Rician and Rayleigh distributions are depicted in Figure 2.3.

Example realizations of channel impulse responses and corresponding frequency responses are depicted in Figure 2.4 and Figure 2.5. It can be observed that for a channel with a longer channel impulse response (i.e., a longer τ_{rms}, Figure 2.5), the fading rate in the frequency domain is higher, which means that frequency selectivity is more severe.

The second manifestation of the small-scale fading, besides signal dispersion in the time domain is time-selective fading due to motion. This can be seen as signal

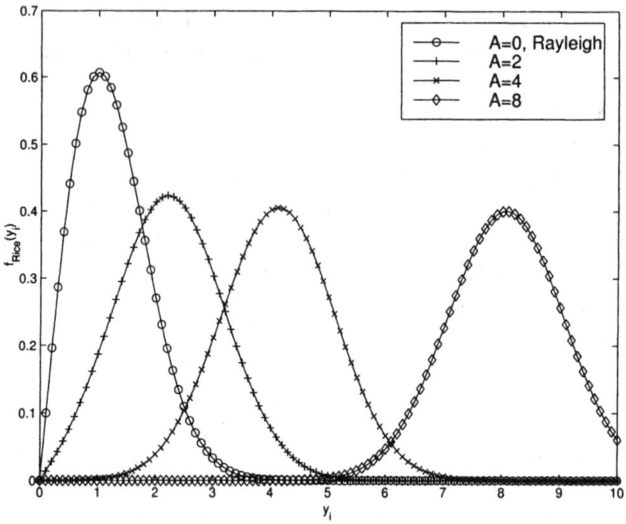

Figure 2.3 Example PDFs of the Rician distribution ($\sigma_s^2 = 1$). A special case of Rayleigh distribution is shown for $A = 0$.

dispersion in the frequency domain commonly known as Doppler spread or Doppler shift. The Doppler frequency that is the frequency observed by the receiver when the receiver and transmitter are in relative motion is defined as [1]

$$f_d = \frac{\vartheta}{\lambda}, \tag{2.14}$$

where ϑ is the relative velocity and λ is the signal wavelength. The measure for time selectivity is the channel's *coherence time*, which can be expressed as [1]

$$T_{coh} \approx \frac{1}{f_d}. \tag{2.15}$$

The coherence time is the time during which the channel response can be considered invariant.

2.1.2 Large-scale effects

Large-scale effects such as path loss and shadowing are responsible for fluctuations of the received signal's power corresponding to large-scale displacement between

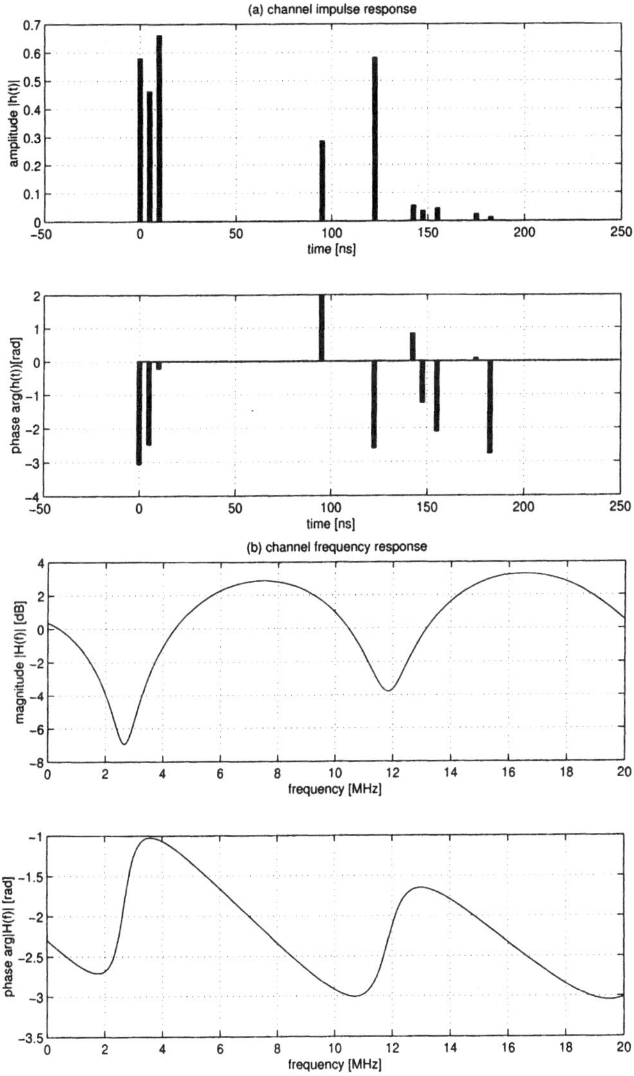

Figure 2.4 Example realization of small-scale fading presented in the time and frequency domain. The following channel parameters are applied: τ_{rms} = 50 ns, $\tau_{max} = 5\tau_{rms}$, K_{factor}=-3 dB.

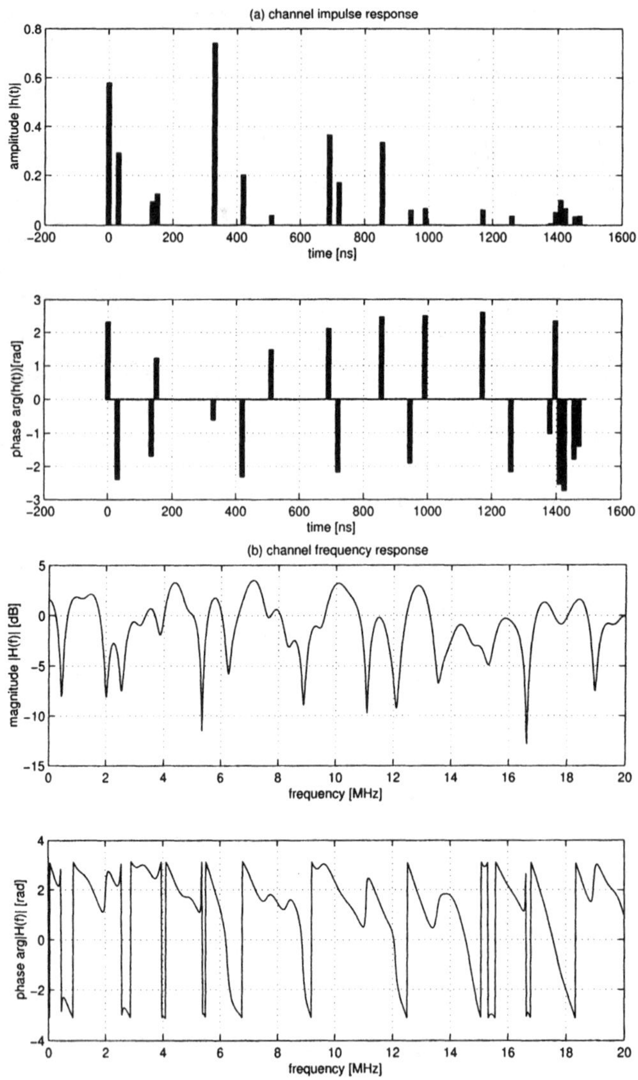

Figure 2.5 Example realization of small-scale fading presented in the time and frequency domain. The following channel parameters are applied: τ_{rms} = 300 ns, $\tau_{max} = 5\tau_{rms}$, K_{factor}=-3 dB.

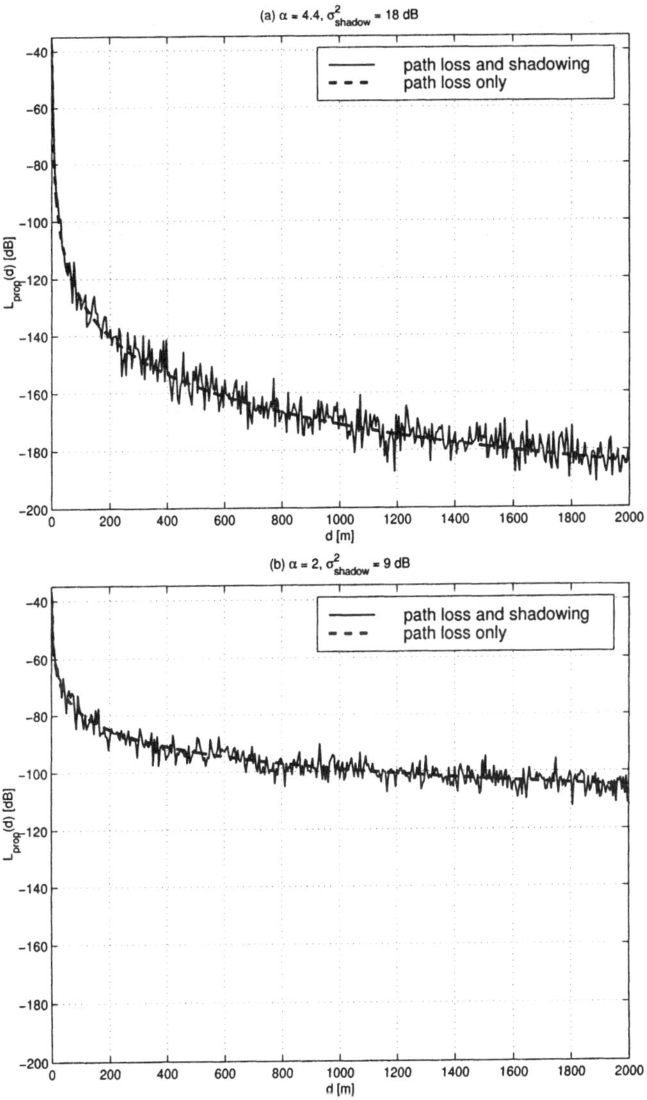

Figure 2.6 Example realizations of large-scale fading (i.e., path loss and shadowing), for various system parameters: α and σ^2_{shadow}.

the transmitter and the receiver. Path loss (PL) depends directly on the distance between the transmitter and the receiver and represents the mean signal attenuation. The following simple model is often used to model the path loss [1]

$$PL(d) = PL(d_0) + 10\alpha \log_{10}\left(\frac{d}{d_0}\right), \tag{2.16}$$

where d_0 is a reference distance, d is the distance between the transmit and receive antennas, and α is the path loss exponent. As a reference distance, often $d_0 = 1$m is chosen. Then, the path loss at the reference distance $d_0 = 1$m can be expressed as

$$PL(\lambda, d_0 = 1m) = 10 \log \left(\frac{4\pi}{\lambda}\right)^2. \tag{2.17}$$

Path loss can be related to the average power response of the multipath channel using the following formula

$$PL = 10 \log \left(\frac{1}{\overline{G}_c}\right) = 10 \log \left(\frac{1}{\sum_{p=1}^{P} \beta_p^2}\right). \tag{2.18}$$

The next large-scale fading phenomenon considered is shadowing, also known as log-normal fading. This is caused by the terrain configuration and signal diffraction over large objects that are placed on the communication path. Shadowing results in signal variations around the (average) path loss. Shadowing can be modelled by a zero-mean, Gaussian random variable X_{shadow} (in decibels, dB) with standard deviation σ_{shadow} (in dB), added to the path loss, (2.16). Then the propagation loss expressed in dB equals to $L_{prop}(d) = PL(d) + X_{shadow}$. The log-normal PDF of X_{shadow} can be expressed as [9]

$$f_{log-normal}(X_{shadow}) = \frac{1}{\sigma_{shadow}\sqrt{2\pi}} \exp\left(-\frac{X_{shadow}^2}{2\sigma_{shadow}^2}\right). \tag{2.19}$$

Example realizations of large scale fading phenomena including path loss and shadowing are shown in Figure 2.6.

2.2 BASICS OF OFDM

One possible way of resolving ISI present in multipath channels is to split the data stream with a high bit rate R into N parallel substreams, each having a lower bit

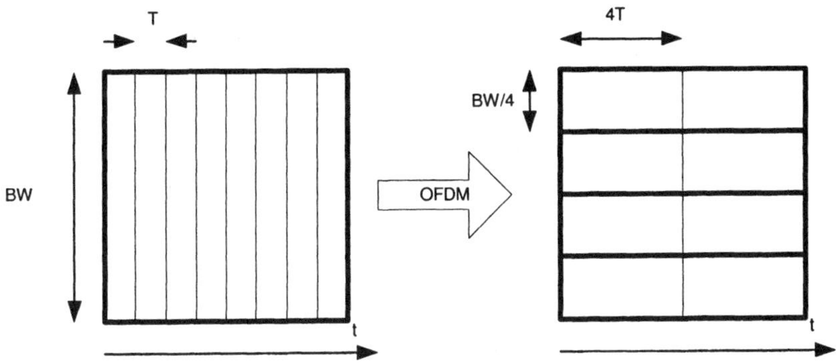

Figure 2.7 Schematic representation of parallel transmission using $N = 4$ subcarriers over total bandwidth of BW.

rate of R/N. Since the substreams are parallel, the symbol duration of each stream is increased and thus the degrading influence of multipath echoes is reduced, as can be observed in Figure 2.7, for the example case of $N = 4$ subcarriers.

The N substreams are modulating N independent carriers, called subcarriers, which are overlapping but orthogonal, as shown in Figure 2.8, and hence distinguishable. This results in a high spectral efficiency. The name for this scheme is orthogonal frequency division multiplexing (OFDM).

OFDM systems belong to the family of multicarrier systems that have been studied since the 1960s. Early multicarrier systems were considered mainly for military applications [11, 12]. A multicarrier system using overlapping but orthogonal carriers (i.e., an OFDM system) was studied in [13] and [14] and was patented in the United States in 1970 [15]. The principal objection to the widespread use of these systems in the past was the complexity of the equipment: a high number of filters and modulators was required for system implementation. In 1971 in [16], Weinstein and Ebert proposed to generate the OFDM signal using the discrete fourier transform (DFT), which greatly reduced the equipment complexity. In such a system, frequency division is achieved not by bandpass filtering but by baseband processing, which can be easily implemented in the digital domain. Another important idea was to apply different modulation levels for different subcarriers according to the subcarriers' quality. The solution proposed by Kalet in [17] is often referred to as "water-filing." It was later followed (among others) in [18] by Steele and Webb, and in [19] by Czylwik. A broader historical perspective on OFDM can be found in [20] or [21].

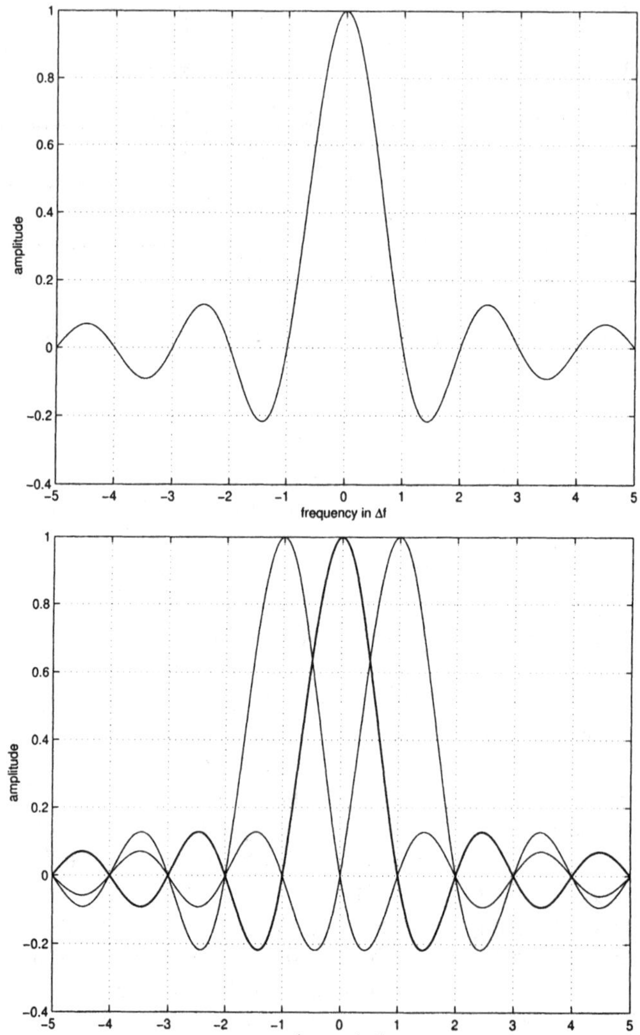

Figure 2.8 Orthogonality between OFDM subcarriers separated by Δf distance. Upper picture shows a single OFDM subcarrier, while lower picture shows three adjacent subcarriers.

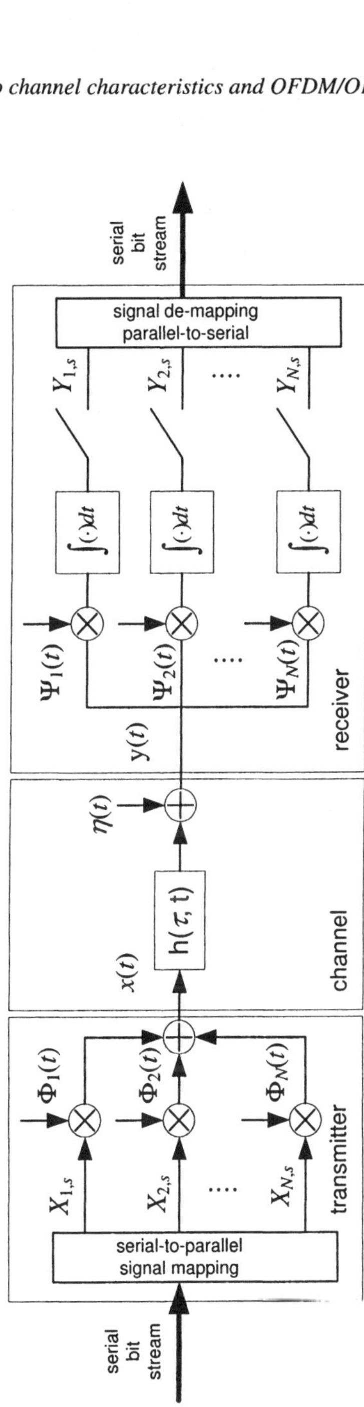

Figure 2.9 Baseband block model of OFDM system.

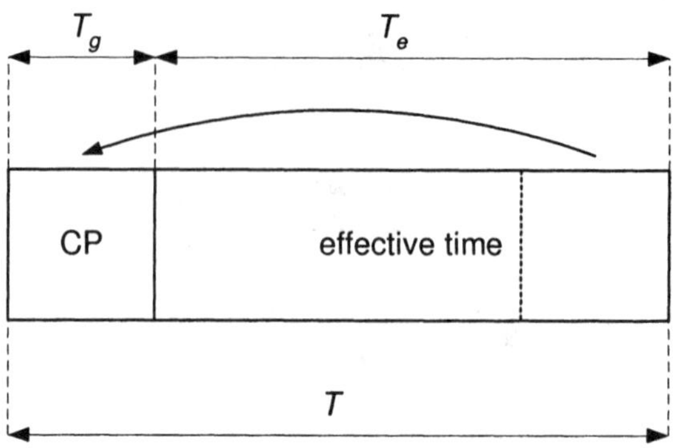

Figure 2.10 Single OFDM symbol with the cyclic prefix (CP) of duration T_g and the effective part of duration T_e.

The baseband model of an OFDM system is depicted in Figure 2.9. At the OFDM transmitter, the serial bit stream with a high bit rate R is split into N parallel substreams, $n = 1, \ldots, N$, each transmitting at a lower bit rate of R/N. After symbol mapping the set of N parallel complex symbols $\{X_{n,s}\}$ modulate N subcarrier waveforms, and after summation, results in the following sequence of transmitted OFDM symbols [5]

$$x(t) = \sum_{s=-\infty}^{\infty} x_s(t - sT), \tag{2.20}$$

where a single OFDM symbol $x_s(t)$ can be expressed by

$$x_s(t) = \sum_{n=1}^{N} X_{n,s}\Phi_n(t), \tag{2.21}$$

with subcarrier waveforms $\Phi_n(t)$ used at the transmitter described by [5]

$$\Phi_n(t) = \begin{cases} \frac{1}{\sqrt{T-T_g}} e^{j2\pi n\Delta f(t-T_g)} & \text{if } 0 \leq t \leq T \\ 0 & \text{otherwise} \end{cases}, \tag{2.22}$$

where T_g is the guard time. The separation between the two neighboring subcarriers Δf is the inverse of the subcarrier's effective symbol time $T_e = T - T_g$, $\Delta f =$

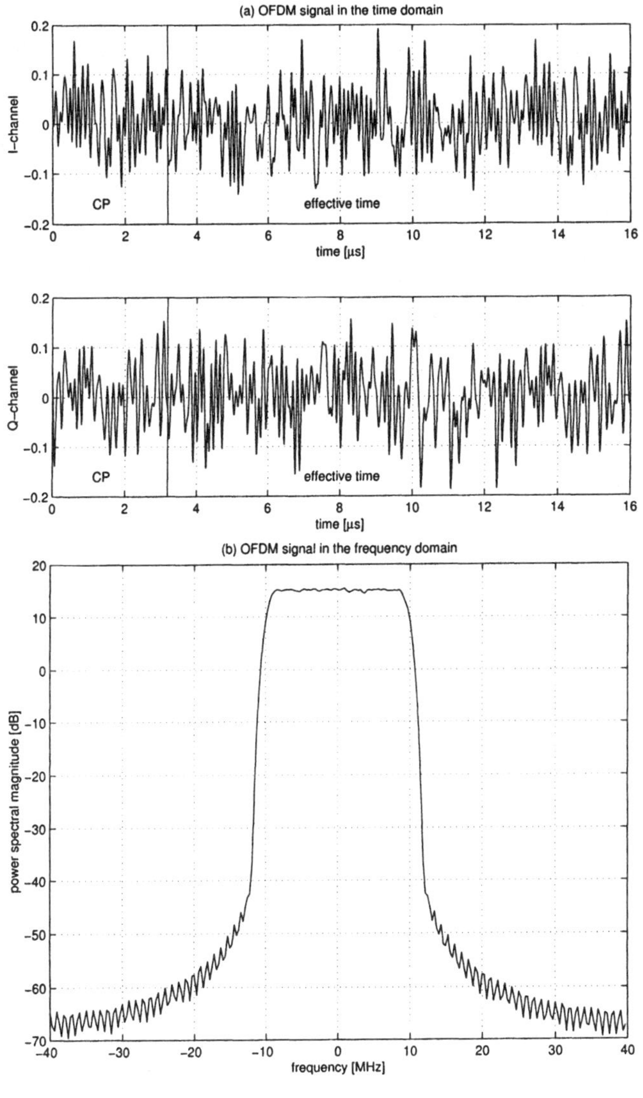

Figure 2.11 Example OFDM signal as seen in (a) the time and (b) the frequency domain, assuming rectangular pulse shaping. N = 256 subcarriers, $T = 16\mu s$, $T_e = 12.8\mu s$, $T_g = 0.2 \cdot T_e = 3.2\mu s$, total bandwidth BW=20 MHz.

$1/T_e$, as can be observed in Figure 2.8. With N subcarriers, the total system bandwidth can be expressed by $BW = N\Delta f$. The guard time T_g is usually filled with a copy of the last part of the OFDM symbol, which allows to eliminate inter-carrier interference (ICI) [22], and is referred to as a cyclic prefix (CP). The duration of the entire OFDM symbol including both effective symbol time and CP duration equals $T = T_e + T_g$, as depicted in Figure 2.10. The duration of the CP is chosen to be longer than the rms delay spread of the channel τ_{rms} in order to shield from ISI, usually $\frac{T_g}{\tau_{rms}} = 2, \ldots, 4$ and $\frac{T_g}{T_e} = 0.2, \ldots, 0.25$ [22]. The values supported in WirelessMAN-OFDM and WirelessMAN-OFDMA air interface are as follows: $\frac{T_g}{T_e} = \{1/32, 1/16, 1/8, 1/4\}$ [23]. An example OFDM signal represented in the time and frequency domains is depicted in Figure 2.11.

The transmitted signal passes through a multipath channel described by the channel impulse response $h(\tau; t)$ and AWGN $\eta(t)$ is added. The received signal can be expressed by

$$y(t) = x(t) * h(\tau; t) + \eta(t) = \int_0^{T_g} h(\tau; t)x(t - \tau)d\tau + \eta(t). \tag{2.23}$$

At the receiver, the signal $y(t)$ passes through a matched-filter bank characterized by the following receiver waveforms [5]

$$\Psi_n(t) = \begin{cases} \Phi_n^*(T - t) & \text{if } 0 \le t \le T - T_g \\ 0 & \text{otherwise} \end{cases}. \tag{2.24}$$

The filter bank is matched to the last part $[T_g, T]$ of the transmitter waveforms $\Phi_n(t)$, which means that the CP is removed at the receiver. Recalling the assumption that the channel impulse response does not exceed T_g, the sampled output from the receiver filter bank is free from ISI. In this way, the negative effect of ISI, already reduced by a lower signaling rate at each subcarrier, is completely eliminated by the use of a CP. The sampled output of the nth matched filter can now be expressed by [5]

$$Y_n = y(t) \otimes \Psi_n(t) = \int_{-\infty}^{\infty} y(t)\Psi_n(T - t)dt. \tag{2.25}$$

Substituting $y(t)$ with (2.23), $\Psi_n(T - t)$ with $\Phi_n^*(t)$, and changing integration interval of the outer integral from $[-\infty, \infty]$ to $[T_g, T]$ leads to the following relation

$$Y_n = \int_{T_g}^{T} \left(\int_0^{T_g} h(\tau; t)x(t - \tau)d\tau + \eta(t) \right) \Phi_n^*(t)dt, \tag{2.26}$$

which, after substituting $x(t-\tau)$ with (2.21) and ignoring the index s (since the sampled output from the receiver filter bank contains no ISI) leads to

$$Y_n = \int_{T_g}^{T} \left(\int_0^{T_g} h(\tau;t) \sum_{\overline{n}=1}^{N} X_{\overline{n}} \Phi_{\overline{n}}(t-\tau) d\tau \right) \Phi_n^*(t) dt + \int_{T_g}^{T} \eta(T-t) \Phi_n^*(t) dt. \tag{2.27}$$

Assume that the channel is time-invariant during the transmission of one OFDM symbol. Then $h(\tau)$ can be used instead of $h(\tau;t)$, which leads to a further simplification

$$Y_n = \sum_{\overline{n}=1}^{N} X_{\overline{n}} \int_{T_g}^{T} \left(\int_0^{T_g} h(\tau) \Phi_{\overline{n}}(t-\tau) d\tau \right) \Phi_n^*(t) dt + \int_{T_g}^{T} \eta(T-t) \Phi_n^*(t) dt. \tag{2.28}$$

The inner integral of (2.28) can be expressed using (2.22) as

$$\int_0^{T_g} h(\tau) \Phi_{\overline{n}}(t-\tau) d\tau = \int_0^{T_g} h(\tau) \frac{e^{j2\pi\overline{n}\Delta f(t-\tau-T_g)}}{\sqrt{T-T_g}} d\tau. \tag{2.29}$$

After a simple reorganization of (2.29), we obtain

$$\int_0^{T_g} h(\tau) \Phi_{\overline{n}}(t-\tau) d\tau = \frac{e^{j2\pi\overline{n}\Delta f(t-T_g)}}{\sqrt{T-T_g}} \int_0^{T_g} h(\tau) e^{-j2\pi\overline{n}\Delta f\tau} d\tau. \tag{2.30}$$

The part of expression (2.30) in front of the integral is equivalent to the transmitter waveform $\Phi_{\overline{n}}(t)$ at subcarrier $\overline{n}$, as given in (2.22). The remaining part of the expression (2.30) (i.e., the integral) can be perceived as the Fourier transform of the channel impulse response $h(\tau)$ sampled at $\overline{n}\Delta f$. In other words, this is the channel frequency response $H_{\overline{n}} = H(\overline{n}\Delta f)$ at frequencies $\overline{n}\Delta f$ [5]

$$H_{\overline{n}} = H(\overline{n}\Delta f) = \int_0^{T_g} h(\tau) e^{-j2\pi\overline{n}\Delta f\tau} d\tau. \tag{2.31}$$

Using the above two transformations in the expression (2.30) leads to

$$\int_0^{T_g} h(\tau) \Phi_{\overline{n}}(t-\tau) d\tau = \Phi_{\overline{n}}(t) H_{\overline{n}}, \tag{2.32}$$

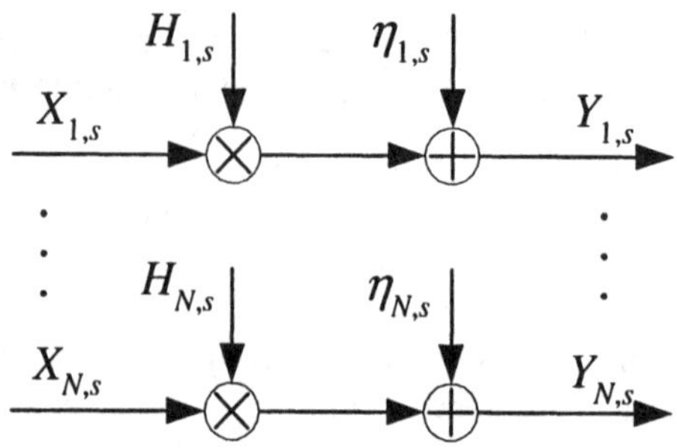

Figure 2.12 Equivalent OFDM system considered as a set of parallel Gaussian channels.

which, after substitution in the last expression presenting the sampled output Y_n from the receiver filter bank (2.28), gives the following relation

$$Y_n = \sum_{\overline{n}=1}^{N} X_{\overline{n}} H_{\overline{n}} \int_{T_g}^{T} \Phi_{\overline{n}}(t)\Phi_n^*(t)dt + \int_{T_g}^{T} \eta(T-t)\Phi_n^*(t)dt. \tag{2.33}$$

Because the transmitter waveforms $\Phi_n(t)$ are orthogonal, the first integral in the above relation equals to one only when $\overline{n} = n$, otherwise the integral equals to zero. The second integral can be explicitly indicated as independent noise samples $\eta_n = \int_{T_g}^{T} \eta(T-t)\Phi_n^*(t)dt$. This leads to the following simplification of (2.33) [5,24]

$$Y_n = X_n H_n + \eta_n. \tag{2.34}$$

This essentially means that a frequency selective channel has been replaced by a set of parallel channels, each of which is flat fading. Additionally, such a model indicates the potential for adapting communication parameters (e.g., modulation, coding) per OFDM subcarrier independently. The equivalent OFDM system is schematically depicted in Figure 2.12.

The process of OFDM modulation is usually realized with the use of inverse discrete fourier transform (IDFT), usually implemented by the computationally efficient inverse fast fourier transform (IFFT), while the process of OFDM demodulation is realized with the use of discrete fourier transform (DFT) usually implemented as fast fourier transform (FFT). The corresponding baseband discrete system model, where IDFT and DFT are used, is depicted in Figure 2.13.

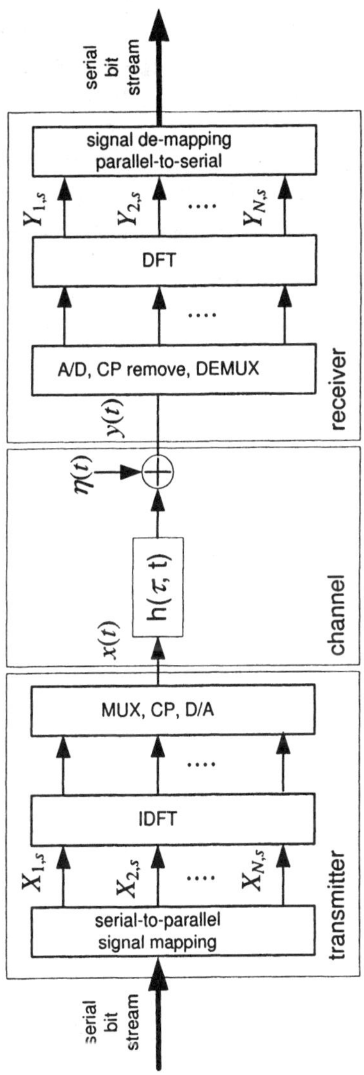

Figure 2.13 Baseband discrete block model of OFDM system applying IDFT and DFT.

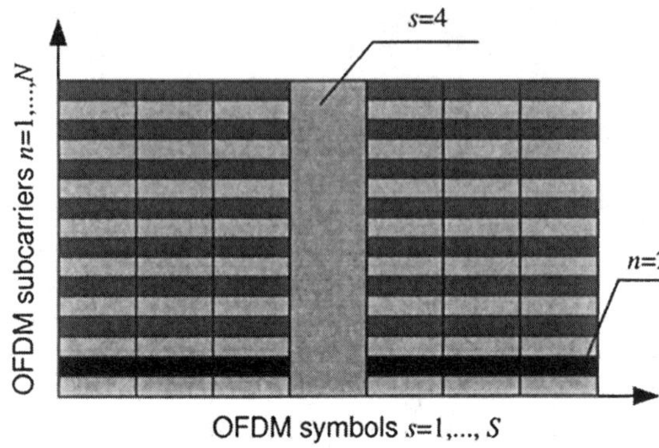

Figure 2.14 OFDM resources: OFDM symbols $s = 1, \ldots, S$ and subcarriers $n = 1, \ldots, N$.

2.3 ORTHOGONAL FREQUENCY DIVISION MULTIPLE ACCESS

In addition to high bandwidth efficiency and robustness to ISI from multipath propagation, OFDM has also other important advantages, which are evident when considering resource management for multiuser communications. It allows for partitioning the time and frequency resources into

- OFDM symbols, $s = 1, \ldots, S$,
- OFDM subcarriers, $n = 1, \ldots, N$.

The schematic representation of this partitioning is shown in Figure 2.14. Each of the OFDM symbols or subcarriers can be assigned to different users. Moreover, different users may be allocated different numbers of symbols and subcarriers, and in addition different levels of modulation (number of bits per modulation symbol) can be applied on each subcarrier. In this way various levels of QoS can be realized. Allocation of different subcarriers to different users for the downlink wireless access system is depicted in Figure 2.15. As is shown in this case, each user is allocated a number of subcarriers exclusively, depending on his or her traffic needs and channel conditions. Which subcarrier is allocated to which user remains the task of the radio resource manager (often known as the scheduler).

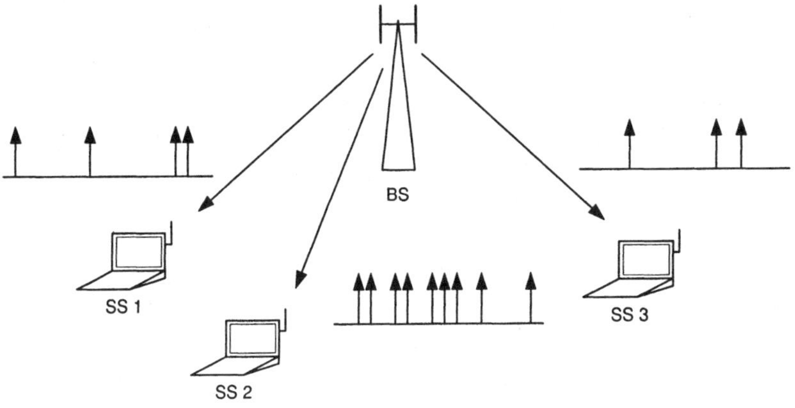

Figure 2.15 Downlink OFDMA transmission addressing different users with different subcarriers.

2.4 DISCUSSION

OFDM exhibits significant advantages as a transmission technology. Below, some of the most important advantages are listed:

- OFDM is robust to multipath propagation thanks to the prolonged duration of symbols on each subcarrier and the application of a cyclic prefix.
- OFDM is easy to implement in the digital domain thanks to the possibility to use (I)FFT chips.
- OFDM provides high resource granularity by splitting the broadband frequency channel into a number of narrowband subcarriers. This granularity is useful from the perspective of QoS support and resource management for satisfying user traffic needs as close as possible with the least possible amount of resources used.
- OFDM is bandwidth efficient, since the parallel subcarriers are overlapping yet orthogonal and thus not causing interference to each other.

As usual, there are also some disadvantages of this technology:

- OFDM suffers from the so-called peak-to-average problem (PAP). The OFDM signal is composed out of N independently modulated signals, which, when added with the same phase, result in an N-times higher output power than the average [22]. Therefore, signal power fluctuations can be very dynamic, which causes problems in the design of RF amplifiers and AD/DA converters.

- OFDM is susceptible to reference carrier frequency shifts, which degrades the orthogonality between subcarriers. Research on efficient synchronization methods for OFDM constitutes a lion's share of OFDM research in the late 1990s.

OFDMA inherits all the advantages and disadvantages of OFDM, and thanks to the multiuser scenario it also exhibits some new features. The main new advantages of OFDMA include:

- OFDMA provides *multiuser diversity*: a subcarrier that is of low quality to one user is usually not of low quality to other users and thus should be allocated accordingly.
- OFDMA is flexible: subcarriers addressed to users can be independently power controlled and a suitable modulation level can be applied on a per-subcarrier basis.

The main disadvantages of OFDMA stem from the fact that the multiuser environment is, by definition, prone to additional cochannel interference. This calls for more advanced methods for resource distribution, which aim at minimization of the negative influence of (and to) the other users. One of the most challenging and interesting topics in radio communication research is radio resource management. In the OFDMA context, radio resource management means allocation of users to base stations, subcarriers to users, as well as determination of modulation levels (bit loading) and power control per individual subcarrier. In this book, novel methods are discussed that are suitable for efficient radio resource management for OFDMA systems, with particular focus on schemes operating in a multiuser interference environment.

References

[1] Gibson, J.D., (ed.), *The Mobile Communications Handbook,* CRC Press, 1999.

[2] Parsons, D., *The Mobile Radio Propagation Channel,* Pentech Press, London, 1992.

[3] Bohdanowicz, A., *Wideband Indoor and Outdoor Radio Channel Measurements at 17 GHz,* final report, Delft University of Technology, February 2000.

[4] Janssen, G.J.M., *Robust receiver techniques for interference-limited radio channels,* Delft University of Technology, PhD Dissertation, 1998.

[5] Edfors, O., et al., *An introduction to orthogonal frequency-division multiplexing,* technical report, Lulea University of Technology, September 1996.

[6] van de Beek, et al., "A Conceptual Study of OFDM-based Multiple Access Systems, Part 1: Air Interface Requirements," ETSI Technical Report, 1996.

[7] Saleh, A.A.M., and Valenzuela, R.A., "A Statistical Model for Indoor Multipath Propagation," *IEEE Journal on Selected Areas in Communications,* February 1987, volume 5, pp. 128-137.

[8] Proakis, J.G., *Digital Communications,* McGraw-Hill, 1995.

[9] Lee, W.C.Y., *Mobile Communication Design Fundamentals,* John Wiley and Sons, 1993.

[10] Lee, W.C.Y., *Mobile Cellular Telecommunications Systems,* McGraw-Hill Book Company, 1989.

[11] Bello, P.A., "Selective Fading Limitations of the KATHRYN Modem and Some System Design Considerations," *IEEE Transactions on Communications Technology,* September 1965.

[12] Zimmermann, M.S., and Kirsch, A.L., "The AN/GSC-10/KATHRYN Variable Rate Data Modem for HF Radio," *IEEE Transactions on Communications Technology,* April 1967, volume 15, pp. 197-205.

[13] Chang, R.W., "Synthesis of Band-Limited Orthogonal Signals for Multichannel Data Transmission," *Bell Syst. Tech. J.,* December 1966, volume 45, pp. 1775-1796.

[14] Saltzberg, B.R., "Performance of an Efficient Parallel Data Transmission System," *IEEE Transactions on Communications Technology,* December 1967, volume 15, pp. 805-811.

[15] *Orthogonal frequency division multiplexing,* U.S. Patent 3,488,445, filed Nov. 14, 1966, issued Jan. 6, 1970.

[16] Weinstein, S.B., and Ebert, P.M., "Data Transmission by Frequency-Division Multiplexing Using the Discrete Fourier Transform," *IEEE Transactions on Communications Technology,* October 1971, volume 19, pp. 628-634.

[17] Kalet, I., "The Multitone Channel," *IEEE Transactions on Communications,* February 1989, volume 37, pp. 119-124.

[18] Steele, R., and Webb, W., "Variable Rate QAM for Data Transmission over Rayleigh Fading Channels," *Proc. Wireless,* Atlanta, pp. 1-14, 1991.

[19] Czylwik, A., "Adaptive OFDM for Wideband Radio Channels," *Proc. IEEE Globecom,* pp. 713-718, 1996.

[20] Keller, T., and Hanzo, L., "Adaptive Multicarrier Modulation: A Convenient Framework for Time-Frequency Processing in Wireless Communications," *IEEE Proceedings of the IEEE,* May 2000, volume 88, pp. 611-640.

[21] Zou, W.Y., and Wu, Y., "COFDM: An Overview," *IEEE Transactions on Broadcasting,* March 1995, volume 41, pp. 1-8.

[22] van Nee, R., and Prasad, R., *OFDM Wireless Multimedia Communications,* Artech House, 2000.

[23] IEEE, *Part 16: Air Interface for Fixed Broadband Wireless Access Systems,* IEEE Std 802.16-2004, October 2004.

[24] Witrisal, K., *OFDM System Proposal,* 2nd progress report, IRCTR, Delft University of Technology, November 1998.

Chapter 3

Radio resource management principles

Wireless access systems are often considered to be the weakest link of the entire communication chain [1]. This is not only due to the difficult and unpredictable behavior of the radio channel but also due to the fact that radio resources are much more limited and scarce than for example cable resources. The problem is that while the demand for more traffic increases at a very high pace, the availability of new spectra increases at a much lower pace. This leads to another phenomenon, which may seem contradictory at first glance but as appears later in this book is real: more traffic should and can be carried by a system without allocating new resources but by making better use of the available ones. This brings us to the three, closely coupled aspects in a multiuser wireless communication system: multiple access, QoS support, and resource management.

3.1 BASICS OF MULTIPLE ACCESS

The main task for multiple access techniques is to allow multiple users to communicate by sharing the finite physical resources of a system [2], such as time, bandwidth, and transmit power. Sharing involves not only creating separable links for multiple communications as depicted in Figure 3.1, but also *reuse* of available resources in order to increase their utilization. *Reuse* means that at the same time instant more than one user is assigned the same resource (e.g., a frequency channel or a time slot). Therefore, these multiple users influence each other, sometimes causing problems, as in the case of cellular systems (as described in Chapter 7) but sometimes bringing benefits, as in the case of embedded modulation systems (as described in Chapter 6). In any case, it is of key importance that this mutual influence,

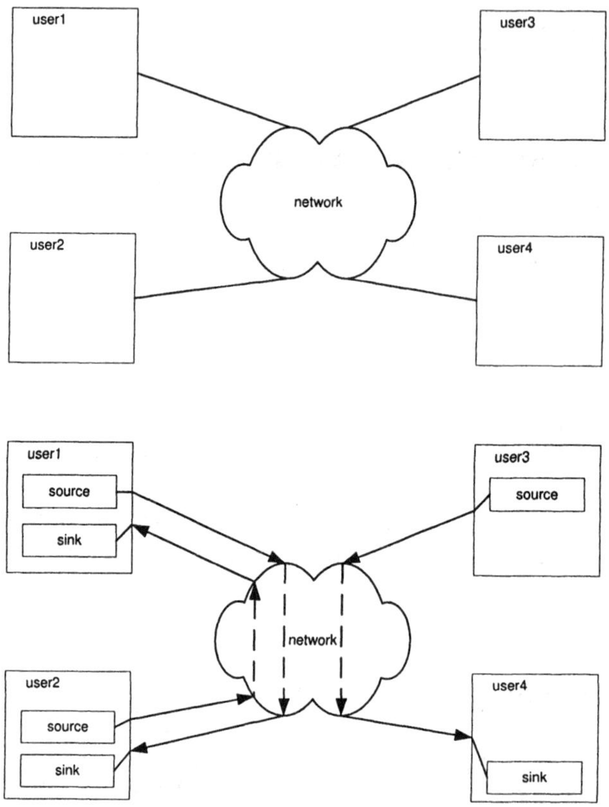

Figure 3.1 Creating separable links for multiple communications.

also referred to as interference, is controlled. As indicated by Viterbi: "Whichever multiple access technique is employed, the ultimate performance limitation is the system's susceptibility to interference" [3].

3.2 THE NEED FOR QOS SUPPORT

The need for supporting various traffic requirements stems from the foreseen multitude of future multimedia applications [4, 5]. However, at this moment, it is difficult to say what will be "killer applications." Therefore, the communication system should be ready to anticipate almost any kind of application types. Having such a reconfigurable system that can support various types of customizable services is beneficial from the time-to-market perspective, where the operator can quickly react to the market needs by tailoring a specific service, without the need for time consuming system development. This is not the case for the majority of existing systems, where a different communication service means the need for a different network. Moreover, differentiation of the users' traffic requirements allows for allocating just as much resources as needed without wasting them in a worst-case design approach. In this way the network resources are well utilized. Therefore, support of various traffic profiles, often referred to as support of QoS, is of prime importance in designing future multiple access schemes.

QoS should be provided in an *end-to-end* way; that is, it should be supported by all system elements that constitute the path between the communicating parties [6]. Often, the wireless link is the weakest link of the communication path due to its variable and unpredictable behavior. This motivates research on finding efficient methods for QoS support in a wireless access system. One of the possibilities for building end-to-end QoS provisioning can be done by negotiating QoS parameters, agreeing on certain parameter settings and establishing service level agreements (SLAs) between the involved system elements [7]. This approach is widely used in standardized systems, such GPRS or UMTS, and is also adopted in this book (a corresponding QoS model is introduced in Chapter 5).

The system providing QoS should be fair. QoS provision for one user should not inhibit QoS provision for another user, assuming users have the same priorities. Also, one user should not have his or her QoS provided at the expense of other users. This rule is often violated in popular "best-effort" multiple-access systems, such as current implementations of GPRS.

3.3 THE NECESSITY FOR RESOURCE MANAGEMENT

In wireless communications, as in real life, when sharing of finite resources to satisfy individual needs (i.e., QoS) is involved, it is always related to management of these resources, since the sharing should be done in a fair and efficient way. The following general approaches to radio resource management can be identified [8,9]:

- *Fixed resource allocation*, which is historically the oldest approach to the problem of resource management. Here, certain resources (such as frequency channels or time slots) are assigned permanently to certain users. This approach is still present in contemporary communication systems, for example in GSM, where one of the options for frequency planning is to assign a frequency permanently to transceivers at certain base stations. This is a worst-case design aiming to provide a minimum carrier-to-interference (C/I) ratio over the majority of a system's service area. Obviously, fixed resource allocation, though simple and not requiring the knowledge of the environment, is a waste of resources for most of the time when the worst-case assumption is too stringent.
- *Random resource allocation*, where resources are either randomly allocated or randomly accessed. Examples of randomly accessed time resource are all multiple access schemes based on ALOHA and its derivatives (CSMA, CSMA/CD). Examples of randomly allocated resources are spread spectrum (SS) techniques, such as frequency hopping (FH) or direct sequence (DS) SS. These SS systems can also be treated as interference averaging techniques [9], and usually they provide better performance than fixed resource allocation. Their advantage stems from the fact that without having a detailed knowledge of the environment, on average, it is better to allocate resources randomly, since the negative effect of the environment (noise, interference, channel attenuation) will be averaged over all users without risking that some of them are significantly affected, as is the case of fixed resource allocation.
- *Dynamic resource allocation*, where, based on a detailed knowledge about the environment and the user's requirements, the resource manager makes decisions about which resource should be allocated to which user. In this way, resources are allocated according to their quality and users' needs. One of the examples of dynamic resource allocation is dynamic channel allocation (DCA) [10, 11], where a channel from a common pool is allocated to a user based on the measured C/I status. Dynamic resource allocation operating in the interference limited environment can also be named as interference avoidance [9]. These methods are claimed in [9] to provide an even better performance

than the interference averaging methods. Here, adaptation is seen as the key success factor [12]. The drawback of these methods is the necessity to obtain detailed information about the environmental conditions.

In this book, the focus is on the dynamic resource allocation techniques, requiring and utilizing the knowledge about the environment in which the system operates. In combination with diversity and feedback [9], they constitute a powerful foundation for building efficient multiple access schemes.

3.4 RADIO RESOURCE MANAGEMENT IN CONTEMPORARY OFDM / OFDMA SYSTEMS

In most of the contemporary systems OFDM is used merely as a transmission scheme. This means that for a certain time slot, all OFDM subcarriers are allocated to one user, regardless their quality. Only in the case of 802.16 specification of the WirelessMAN-OFDMA air interface and for some proprietary solutions, true OFDMA approach is applied, where a user may be assigned a portion of subcarriers. This not only adds to an increased resource granularity but also opens up the possibility for adaptive allocation of subcarriers to users according to the user-perceived channel quality. However, the subcarriers constituting a subchannel in WirelessMAN-OFDMA are selected in a pseudo-random process in order to provide diversity and randomize channel quality and interference. As indicated in [13], further performance gains could be expected if adaptive subcarrier allocation is applied.

Regarding other radio resource management mechanisms, some systems (i.e., HiperLAN/2, 802.11h, 802.16) allow for:

- Dynamic frequency selection that is fitting the entire system band into the most appropriate place of the available spectrum.
- Power control, which is used to limit interference to other systems and protect link quality degradation caused by changing propagation conditions.
- Adaptation of modulation modes according to changing C/I.

These possibilities still do not allow to fully exploit the potentials present in the OFDMA technology. First, most of these systems do not utilize multiuser diversity. Second, transmit power control, if applied, is used to protect the quality of a particular link regardless of the other cochannel links. This means that the change of power at a particular link may cause quality degradation to all the other cochannel links. And third, for most of the systems, adaptation (of modulation modes and power levels) is done per entire block of spectrum and not per individual subcarrier or a group of subcarriers. Also, the notion of subcarrier reuse within a cell is addressed only in the DVB-T standard under the name of multiresolution QAM with the application to transmit two resolutions of the same video signal. In none of the discussed systems is it used in a multiple-access scenario. All these reasons open up a broad area for opportunities for further research work and prospective standards' improvements, part of which is addressed in this book.

3.5 RADIO RESOURCE MANAGEMENT FOR OFDMA ADDRESSED IN THIS BOOK

Radio resource management (RRM) can be boiled down to the determination of methods for allocation of subcarriers to users, as well as setting the transmit power level and number of bits per each subcarrier (bit loading). Additionally, in a cellular environment assignment of users to base stations should be performed. These tasks fit well into the tasks of the resource manager as presented in [11], which are as follows:

- Waveform (channel) assignment: in the case of OFDMA system it means subcarrier assignment as well as determination of the modulation level (bit loading) per subcarrier.
- Base station (access port) assignment.
- Transmit power assignment, also referred to as power control.

Although radio resource management for OFDMA systems is a relatively new research area, it has already gained significant attention of the research community. In the following, three OFDMA system cases are introduced. In the first case subcarriers are exclusively allocated to users in a single cell. In the second case subcarriers are reused within the same cell (intracell reuse), while in the third case they are reused between different cells (intercell reuse in a cellular system).

3.5.1 OFDMA without spectrum reuse

A subcarrier that is of low quality to one user may be of high quality to another user and can be allocated accordingly. This is the basic idea exploiting the presence of *multiuser diversity*. In the case of OFDMA systems with exclusive assignment of subcarriers to users, a subcarrier can only be allocated to one user. This simplifies the allocation process, since no cochannel interference has to be taken in the allocation process.

3.5.2 OFDMA with intracell spectrum reuse

A subcarrier can not only be of high quality to one user while being of low quality to another user but it can also be of high quality to a number of users. Thus it may be power efficient to reuse this subcarrier within a cell and stack multiple signals intended for different users on a subcarrier. In this way, an embedded constellation is created, where symbols of one link are centered around the symbols of another link. Application of embedded modulation allows to increase system capacity, since stacking up to L independent (QAM modulated) signals on each of the N OFDM subcarriers creates $L \cdot N$ independent links, compared to at most N links when no stacking is allowed. This can also be perceived as a generalization of adaptive modulation, where bits loaded on a single subcarrier are not necessarily addressed to one user but may be addressed to various users, depending on the underlying channel conditions. Thanks to that, embedded modulation is more power efficient than conventional adaptive modulation. The individual links can be distinguished at the receiver by the application of successive detection and subtraction.

3.5.3 OFDMA with intercell spectrum reuse

The second interference scenario considers a cellular OFDMA system, where subcarriers are dynamically reused among different cells while being exclusively assigned to users within each cell. The cellular system is a basic prerequisite to have OFDMA applied widely as a technique for the next generation networks [14]. In such a system, a reuse factor of one is applied; that is, the entire bandwidth is available in every cell and the selection of a particular subcarrier is subject to local load and channel conditions. In addition to providing flexibility in subcarrier assignment, the OFDMA ability to partition the frequency resource enables application of power control on a per-subcarrier basis in order to provide and maintain the QoS of all the cochannel links. Further in the book, the focus is placed on the aspects related to resource partitioning in the frequency domain solely and thus resource distribution in the time domain is not considered.

References

[1] Verdu, S., "Wireless Bandwidth in the Making," *IEEE Communications Magazine,* July 2000, pp. 53-58.

[2] Arnbak, J.C., "The Europen (R)evolution of Wireless Digital Networks," *IEEE Communications Magazine,* September 1993, pp. 74-82.

[3] Viterbi, A.J., "Wireless Digital Communications: A View Based on Three Lessons Learned," *IEEE Communications Magazine,* September 1991, pp. 33-36.

[4] Ohmori, S. , Yamao, Y., and Nakajima, N., "The Future Generations of Mobile Communications Based on Broadband Access Technologies,", *IEEE Communications Magazine,* December 2000, 134-142.

[5] Lee, C., *On Quality of Service Management,* Carnegie Mellon University, Ph.D. Dissertation, August 1999.

[6] Aurrecoechea, C., Campbell, A.T., and Hauw, L., "A Survey of QoS Architectures," *Multimedia Systems,* Springer-Verlag", 1998, volume 6, pp. 138-151.

[7] Desoblin, G., and Papini, H., "SLA Management: A Key Differentiator for Service Providers," *Alcatel Telecommunications Review,* 3Q 2001.

[8] Kleinrock, L., "On Some Principles of Nomadic Computing and Multi-Access Communications," *IEEE Communications Magazine,* July 2000, 46-50.

[9] Pottie, G.J., "System Design Choices in Personal Communications," *IEEE Personal Communications,* October 1995, volume 2, pp. 50-67.

[10] Katzela, I., and Naghshineh, M., "Channel Assignment Schemes for Cellular Mobile Telecommunication Systems: A Comprehensive Survey," *IEEE Personal Communications,* June 1996, pp. 10-31.

[11] Zander, J., "Radio Resource Management in Future Wireless Networks: Requirements and Limitations," *IEEE Communications Magazine,* August 1997, volume 35, pp. 30-36.

[12] Nanda, S., Balachandran, K., and Kumar, S., "Adaptation Techniques in Wireless Packet Data Services," *IEEE Communications Magazine,* January 2000, pp. 54-64.

[13] Ghosh, A., et al., "Broadband Wireless Access with WiMAX/802.16: Current Performance Benchamarks and Future Potential," *IEEE Communications Magazine,* February 2005, pp. 129-136.

[14] Lawrey, E. P., *Adaptive Techniques for Multiser OFDM,* James Cook University, Ph.D. Dissertation, 2001.

Chapter 4

Overview of the IEEE 802.16 standard

From all the BWA systems currently considered, IEEE 802.16 seems to attract most of the market attention. The IEEE 802.16 is a family of standards defining the PHY and MAC layers for the wireless metropolitan area network (MAN) and is often referred to as WiMAX. The WiMAX system is designed to offer multimedia services with capacities of up to 40 Mbit/s per channel over the cell radius between 3 and 10 kilometers [1]. The 802.16 standards describe various modes of system operation, including specifications for NLOS and LOS environments, as well as for both license and license-exempt frequency bands. This is reflected in multiple specifications for the PHY layer.

The WiMAX system is supported by the WiMAX Forum [1], which is an industry corporation gathering major operators, equipment providers, and component manufacturers. The WiMAX Forum's goal is "to promote and certify the compatibility and interoperability of broadband wireless access equipment that conforms to the IEEE 802.16 and ETSI HiperMAN standards" [1]. This is realized by involving more than 310 companies in the Forum and participating actively in the standardization efforts. A high number of players covering an entire value chain of system production and implementation allows for specialization and healthy competition, which should lead to price reductions and performance increase. WiMAX Forum puts emphasis on its certification program. The idea is to test the interoperability of the WiMAX equipment in order to give the "WiMAX Forum Certified ™" designation to those products that pass the conformance tests, which are carried out by a selected laboratory. It is also worth mentioning that 802.16 brings not only the specification of the system itself but also the specification of the conformance test cases that need to be passed in order to have a system compliant with the standard. Thanks to that, prospective problems stemming from different interpretations of a standard by various vendors can be avoided. Standardization together with certification gives the promise of a truly worldwide and interpretable system. It also adds to the economies of scale making WiMAX systems cost-effective solutions.

In the following subsections, after outlining the 802.16 standard evolution, we will characterize PHY and MAC layers. Then, a presentation of the key 802.16 system profiles will be given and a summary of the 802.16 standard will close this section.

4.1 802.16 STANDARD EVOLUTION

The need for a truly worldwide wireless access standard crystalized in August 1998 during the meeting called by the National Wireless Electronics Systems Testbed (N-WEST) of the U.S. National Institute of Standards and Technology. The initiative was then passed to IEEE 802, which created the 802.16 Working Group on Broadband Wireless Access (BWA) operating since July 1999 [1–3]. The initial 802.16 standard defined PHY and MAC for licensed bands in the frequency range of 10-66 GHz [4]. Such frequencies require line-of-sight (LOS) operation, which limits the system's applicability. Moreover, licensed bands are available only for operators and service providers able to pay the license price. Usually, the license fees can constitute a prohibitively high entrance barrier for smaller players. Thus, the 802.16 Working Group developed the 802.16a [5] amendment approved in January 2003, which extended the previous standard by addressing sub-11 GHz bands allowing for non-LOS (NLOS) operation, which greatly adds to increasing the base station range. In addition, both licensed and license-exempt bands are considered. However, the allowed NLOS operation results in a much more difficult radio condition (more multipath) and therefore methods to combat this adverse effect had to be developed. These methods, also standardized in 802.16a, include OFDM and OFDMA, extending in this way the set of available PHY layers. Both OFDM and OFDMA PHY layers are extremely robust to multipath propagation and, in addition, they expose other very useful features, such as bandwidth efficiency and flexibility for resource allocation in the frequency domain.

The next release of 802.16, namely 802.16d was approved in June 2004 and is currently known as 802.16-2004 [3, 6]. The 802.16-2004 consolidates previous releases and extends them by for example providing some enhancements to the uplink and improving indoor SS operation. It is expected that the first commercial equipment will be based on 802.16-2004 [3], applying WirelessMAN-OFDM,

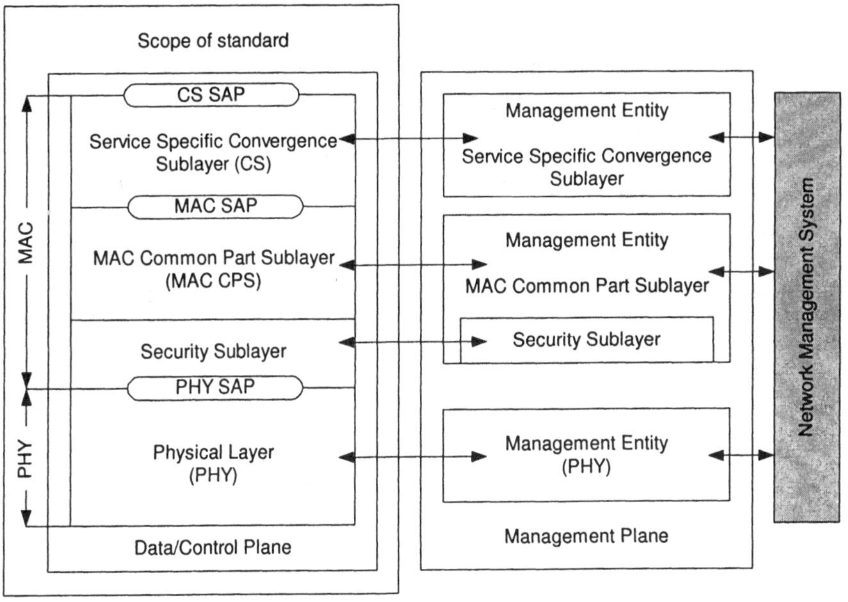

Figure 4.1 IEEE 802.16 system reference model [6]. From IEEE Std. 802.16-2004 Copyright ©2004 IEEE. All rights reserved.

which poses less stringent requirements for implementation than WirelessMAN-OFDMA air interface. The next step in WiMAX standardization is 802.16e, which addresses for mobility support [1, 7].

Below, we summarize the major milestones in the WiMAX system development:

- July 1999 - start of operation of the IEEE 802.16 Working Group
- December 2001 - approval of the early 802.16 standard - LOS operation, licensed bands 10-66 GHz [1].
- January 2003 - approval of 802.16a - sub-11 GHz, both licensed and license-exempt [3].
- June 2004 - approval of 802.16d known as 802.16-2004 [3].
- 802.16e - mobile.

Table 4.1

Air Interfaces Standardized in IEEE 802.16. AAS - Adaptive Antenna System, ARQ - Automatic Repeat Request, STC - Space-Time Coding [6]. From IEEE Std. 802.16-2004 Copyright ©2004 IEEE. All rights reserved.

Air Interface	Band	Additional MAC Features	Duplexing
WirelessMAN-SC	10-66 GHz		TDD, FDD
WirelessMAN-SCa	sub-11 GHz	AAS, ARQ, STC	TDD, FDD
WirelessMAN-OFDM	sub-11 GHz	AAS, ARQ, STC, Mesh	TDD, FDD
WirelessMAN-OFDMA	sub-11 GHz	AAS, ARQ, STC	TDD, FDD
WirelessHUMAN	sub-11 GHz	AAS, ARQ, STC, Mesh	TDD

In the following sections, we provide a description of PHY and MAC layers based on publicly available 802.16-2004 [6], which defines both PMP (point-to-multipoint) and mesh network structure. The system reference model is depicted in Figure 4.1. We begin with the characterization of the PHY layer.

4.2 PHYSICAL LAYER

The PHY layer is the lowest system layer in a hierarchy depicted in Figure 4.1. It communicates with the MAC layer via the PHY layer service access point (SAP). In the 802.16 standard there is one specification of the MAC layer, while there is a number of possible underlying PHY layers, each optimized for different environment. These layers, also referred to as air interfaces, are summarized in Table 4.1. In the following we shall describe each of the standardized air interfaces separately.

4.2.1 WirelessMAN-SC

This PHY layer is based on single-carrier transmission and is aimed for operation in the 10-66 GHz frequency band. It requires LOS between the SS and BS and it is not robust to multipath propagation. The exact mode of the air interface can be adapted to the radio channel conditions and it includes adaptation of modulation and coding on a frame-by-frame basis. The allowed modulation schemes for both uplink and downlink are QPSK, 16-QAM, or 64-QAM (optional). After modulation, baseband pulse shaping using square-root raised cosine filters is applied to the transmitted waveforms. The PHY layer can support both FDD and TDD. When FDD is used the system can serve both full duplex or half duplex SSs. The uplink is based on TDMA, while the downlink is TDD. The air interface also specifies such PHY functions as bit randomization, FEC encoding, synchronization, and frequency control. Uplink power control, compensating free-space power loss or power fluctuations, is part of the standard but details of implementation are left unspecified.

4.2.2 WirelessMAN-SCa

The specification for WirelessMAN-SCa is similar to the one for WirelessMAN-SC. Both are based on a single-carrier technology, both support TDD and FDD and both have TDMA in uplink while having TDM or TDMA in downlink. However, the main difference is that the WirelessMAN-SCa is designed for frequencies below 11 GHz thus for NLOS operation. The multipath propagation, which is evidently present in such environments is handled by the use of channel estimation and equalization. The PHY layer specification allows for adaptive modulation and coding as in the previously described PHY layer but the range of supported modulation schemes is wider. The system supports the following modulation (mapping) methods: spread BPSK (where bits from FEC encoder are spread by a PN sequence prior to BPSK constellation mapper), BPSK, QPSK, 16-QAM, 64-QAM, and finally 256-QAM (optional). A couple of new interesting optional features are introduced when compared to WirelessMAN-SC, including space-time coding (STC) transmit diversity, adaptive antenna support (AAS), and convolutional turbo code.

4.2.3 WirelessMAN-OFDM

As it appears in practice, frequencies below 11 GHz are providing better indoor and outdoor coverage than their counterparts from 10-66 GHz bands. However, multipath propagation is then one of the major impairments limiting the system's performance. To reduce its negative influence, another PHY layer specification was proposed for the 802.16 standard, namely WirelessMAN-OFDM. This PHY layer is designed for NLOS operation, where resulting multipath is overcome by the use of OFDM. In this respect, the specification is similar to WLAN IEEE 802.11a [8], where OFDM is used for the same purpose. For WirelessMAN-OFDM, the transmitter waveforms are generated using a 256-point inverse fast fourier transform (IFFT). This gives in total 256 OFDM subcarriers, out of which 200 are used (either for data or for pilots) and the rest is not active (i.e., used for frequency guards). Also in case of this air interface, adaptive modulation and coding are allowed, where the set of supported modulation schemes includes BPSK, QPSK, 16-QAM, and 64-QAM (optional for license-exempt systems). The FEC can be implemented as concatenated Reed-Solomon convolutional coding or, optionally, block turbo coding or convolutional turbo coding. In license-exempt bands duplexing is based on TDD, while in licensed bands it can be either TDD or FDD. The notion of subchannelization is introduced in the specification of this PHY layer, where a subchannel is a set of OFDM subcarriers. Combination of OFDM subchannel in the frequency domain and OFDM symbols in the time domain form a data block, which can be used as a base for single resource allocation. The SS may or may not support subchannelization. When subchannelization is employed, the transmit power per SS should be proportional to the number of allocated subchannels. Also here, vendors are free to implement their own proprietary power control algorithms.

An interesting optional transmit diversity feature is Alamouti space time coding (STC), which can be used in the downlink. In Alamouti STC, the BS is equipped with two separate OFDM transmitter branches with two antennas transmitting two different OFDM symbols at the same time via two different channels. The SS has a single antenna. At the SS receiver, the estimation of the two channels is performed and this information is used for signal combining and subsequent detection [9, 10]. The PHY layer also provides such classical functions as randomization, interleaving, and synchronization.

4.2.4 WirelessMAN-OFDMA

Yet another PHY layer specified for operation in NLOS environments of below 11 GHz bands is WirelessMAN-OFDMA. OFDMA stands for orthogonal frequency division multiple access, which in addition to reducing the negative effect of multipath propagation provides a basis for distinguishing users in frequency domain, thus allowing for multiple access. Also here, the OFDMA waveforms are generated by the use of IFFT as in the case of WirelessMAN-OFDM. OFDMA subcarriers can carry data, pilots or can be null (used e.g., as guard subcarriers). The total number of OFDMA subcarriers is 2,048, out of which 1,681 are used subcarriers (for data or pilots) and the remaining are not active. Subcarriers are grouped into subsets called subchannels, where each subchannel is made out of 48 subcarriers. Subcarriers in a subchannel can be adjacent (then a so-called adjacent subcarrier permutation is applied), which is useful for support of AAS and AMC or nonadjacent (what can be referred to as distributed subcarrier permutation), which provides extra frequency diversity. MAC data is mapped on OFDMA data region, which is a combination of subchannels in the frequency domain and OFDMA symbols in the time domain.

Transmit power should be proportional to the number of allocated subchannels. Chapters 5, 6, and 7 address various resource allocation problems for BWA systems based on OFDMA and provide results of numerical experiments evaluating their performance.

In WirelessMAN-OFDMA duplexing in licensed bands can be either TDD or FDD, while in license-exempt it can be only TDD. Adaptive modulation and coding are supported, where possible modulation constellations include QPSK, 16-QAM, and 64-QAM. The chosen constellations are then mapped on allocated data subcarriers. Modulation and coding can be adapted on per-allocation bases for downlink and on per-SS bases for uplink. FEC can be based on convolutional coding or, optionally, on block turbo coding or convolutional turbo coding. Additionally, the specification of OFDMA PHY layer allows for support of multiple-input multiple-output (MIMO) techniques, Alamouti STC, and H-ARQ.

4.2.5 WirelessHUMAN

Wireless High Speed Unlicensed Metropolitan Area Network (WirelessHUMAN) is the name used to explicitly indicate 802.16 air interfaces used for unlicensed bands. The air interfaces that can operate in the unlicensed spectra include WirelessMAN-SCa, WirelessMAN-OFDM, and WirelessMAN-OFDMA. Channel bandwidths for WirelessHUMAN are made compatible with 802.11a [8] and are equal to either 10 MHz or 20 MHz.

4.3 MEDIUM ACCESS CONTROL LAYER

The MAC layer is common for all PHY layers in 802.16d [6]. It consists of the three sublayers as can be seen in Figure 4.1, namely

- Security sublayer, which is responsible for security issues such as authentication, encryption, or security key exchange;
- MAC common part sublayer, which is responsible for the main MAC functionalities such as admission control, resource (bandwidth) allocation, and QoS management;
- Service-specific convergence sublayer (CS), which transforms, maps, and adapts external network service data units (SDUs) into MAC SDUs.

In the following, each sublayer is discussed.

Table 4.2
Scheduling Services and Mandatory QoS Flow Parameters [6].

Scheduling service	Mandatory QoS flow parameters
UGS	Maximum Sustained Traffic Rate, Maximum Latency, Tolerated Jitter, Request/Transmission Policy
rtPS	Minimum Reserved Traffic Rate, Maximum Sustained Traffic Rate, Maximum Latency, Request/Transmission Policy
nrtPS	Minimum Reserved Traffic Rate, Maximum Sustained Traffic Rate, Traffic Priority, Request/Transmission Policy
BE	Maximum Sustained Traffic Rate, Traffic Priority, Request/Transmission Policy

4.3.1 Security sublayer

Security sublayer provides security mechanisms for protecting information sent over air interface from interception or service theft. This includes authentication procedures, key management protocol, and data encryption.

4.3.2 MAC common part sublayer

The MAC common part sublayer is central in the entire MAC hierarchy and is responsible for the key MAC functions such as resource allocation, connection establishment, maintenance, and termination. There are two distinct network topologies requiring different handling by the MAC, namely point-to-multipoint (PMP) and mesh topology.

PMP is a topology used in the downlink, where a base station (BS) communicates with users' terminals, also referred to as subscriber stations (SS). In the downlink, either it is indicated that a given part of subframe is addressed for a particular SS or all the SSs listen to the same information, which is then filtered by each SS to retain only the addressed parts. Each SS is identified by a 48-bit MAC address. The BS does not have the need to coordinate its transmissions with other BSs, except for the synchronization for time division duplex (TDD), when it is applied. The MAC protocol is connection oriented. The connection is described by the QoS parameters, which allow for efficient connection management, including connection, establishment, maintenance and termination. Each connection is identified by a 16-bit connection identifier (CID).

In the mesh network, some nodes can be superior to others (BS) or they can be of equal rank. Each node has a 48-bit MAC address. Transmission may be allowed not only between the BS and SSs but also directly between SSs and through SSs. This means that a message from a sender to the recipient can be routed through several radio hops. Stations able to communicate with each other directly are called neighbors. The group of neighbors is called a neighborhood. Scheduling of the transmission between nodes can be distributed or centralized. When a distributed scheduling is used then nodes belonging to a two-hop neighborhood coordinate their transmission, while scheduling information is broadcasted to all the neighbors. In distributed scheduling, downlink and uplink are not differentiated in scheduling mechanisms. In centralized scheduling it is the mesh BS that takes decisions about resource allocation to links after listening to all mesh SS placed within a certain range from the mesh BS. In mesh topology the communication is link-based, where a link is a connection between the two nodes. QoS parameters describe each unicast message belonging to a link. Links in the local neighborhood are identified by using 8-bit link identifier (link ID), which is part of the CID.

The MAC PDU for both PMP and mesh topology consists of a generic MAC header, a variable-length payload, and a cyclic redundancy check (CRC). Depending on the needs, several MAC PDUs can be concatenated into a single transmission or a single MAC service data unit (SDU) can fragmented into several MAC PDUs. Also, packing is possible, where several MAC SDUs are packed into a single MAC PDU. Optionally, ARQ mechanisms can be implemented.

The MAC scheduler supports the following scheduling services:

- Unsolicited grant service (USG), which is applicable to real-time data streams, where fixed-size packets are generated at regular intervals, as for example in T1/E1 links.
- Real-time polling service (rtPS), which is applicable to real-time data streams, where variable-size packets are generated at regular intervals, as for example in MPEG-encoded video.
- Nonreal-time polling service (nrtPS), which is designed for delay-tolerant data streams, where variable-size packets are generated with minimum data rate requirement, as for example in File Transfer Protocol (FTP).
- Best effort (BE), wherc there is no minimum QoS requirement and therefore traffic can be handled in a best-effort manner.

The key point in QoS management is that MAC PDUs are associated with unidirectional service flows, each described by the set of QoS parameters. The relation between each scheduling service and the required QoS flow parameters is depicted in Table 4.2. However, any service class can be configured by providing a set of QoS parameters and their values. The meaning of most useful QoS parameters describing a service flow is the following:

- Traffic priority - is a priority assigned to a service flow. This parameter allows to decide which service flow should be given a preference in the case when all the other QoS parameters of conflicting flows have the same values.
- Maximum sustained traffic rate - is a peak traffic rate of the service expressed in [bit/s] and not including MAC overheads.
- Maximum traffic burst - is a maximum size of a burst for a given service.
- Minimum reserved traffic rate - is a minimum data rate allocated for a service flow and is expressed in [bit/s].
- Minimum tolerable traffic rate - is a minimum rate expressed in [bit/s] that can be tolerated over a certain period of time under additional traffic demand.
- Vendor-specific QoS parameters - allows to implement vendor-specific QoS.
- Service flow scheduling type - indicates the type of used scheduling service as listed in Table 4.2.
- Request/transmission policy - this parameter indicates attributes for a service flow such as whether a service flow shall fragment data or suppress payload headers, or include CRC in MAC PDU.
- Tolerated jitter - is a delay variation in [ms] that can be tolerated by a connection.
- Maximum latency - is a maximum time in [ms] between receiving a packet on the network interface and sending it on the radio interface.

The 802.16 MAC can support both FDD and TDD. In FDD uplink and downlink are organized using different frequencies. The SS can be either full-duplex, when it can simultaneously transmit and receive data or half-duplex, when SS can either transmit or receive at a given instant. In TDD uplink and downlink are usually located on the same frequency and they are separated by different time of transmission. The TDD frame consists of uplink and downlink subframes. The length of these subframes can be adapted according to the needs (usually, downlink requires more resources).

Table 4.3

MAC and PHY Profiles for WirelessMAN-OFDM and WirlessHUMAN (-OFDM). (*From:* IEEE Std. 802.16-2004 [6].)

Id	MAC/PHY	Key feature
profM3_PMP	MAC	PMP operation
profM3_Mesh	MAC	Mesh operation
profP3_1.75	PHY	1.75-MHz channel, licensed
profP3_3.5	PHY	3.5-MHz channel, licensed
profP3_7	PHY	7-MHz channel, licensed
profP3_3	PHY	3-MHz channel, licensed
profP3_5.5	PHY	5.5-MHz channel, licensed
profP3_10	PHY	10-MHz channel, unlicensed

Optionally, for WirelessMAN-SCa, WirelessMAN-OFDM, and WirelessMAN-OFDMA, the 802.16 MAC can support adaptive antenna systems (AAS), as indicated in Table 4.1. The benefits of AAS are manyfold but they can be boiled down to the following two major advantages:

- Capacity increase thanks to the possibility to reuse spectrum much more aggressively, since different links can now be distinguished in the space domain. This applies both to intercell reuse of spectrum, where reuse of one is desired and to intracell reuse of spectrum. Capacity increase is also realized by the ability to limit the cochannel interference, which is accomplished by steering nulls of antenna radiation pattern in the direction of interferers.
- Coverage (range) improvement thanks to the antennas' directivity and possibility to combine signals coming from multiple antennas.

An interesting MAC functionality specified as mandatory for license-exempt operation (WirelessHUMAN) is dynamic frequency selection (DFS). This feature allows to detect and avoid jamming primary users. The SS or BS shall perform channel measurements in order to check the existence of primary users. If such are detected in a given channel, this channel should not be used by a searching SS or BS in order to avoid collision. The BS may instead choose a new channel to operate in. The algorithms for new channel selection are not standardized and are left open for vendor-specific implementations.

The 802.16 MAC supports optionally another beneficial feature, which is hybrid automatic repeat request (H-ARQ). This feature is possible only with

WirelessMAN-OFDMA air interface and it operates as follows. H-ARQ packet is formed by concatenating one or more MAC PDUs. Then, four subpackets are generated out of the PHY burst created from a single H-ARQ packet. In the case of unsuccessful detection of a single subpacket, the subsequent subpackets are sent until the message is correctly received or all four subpackets are transmitted. The chances for correct detection increase with the number of transmitted subpackets since there is redundancy among subpackets and the receiver uses all received subpackets to detect the sent message.

4.3.3 Service-specific convergence sublayer

The service-specific convergence sublayer interconnects MAC common part sublayer (MAC CPS) with higher layer protocols via convergence sublayer (CS) SAP. There are two CS specifications in the 802.16-2004, each interconnecting with different higher layer protocol, namely: asynchronous transfer mode (ATM) CS and packet CS used for all packet protocols such as Internet Protocol (IP), or Ethernet.

The ATM CS allows to connect the 802.16 system to the ATM layer of an ATM network. The system supports both Virtual Path (VP) and Virtual Channel (VC) switching. In both modes, QoS parameters of the ATM connection are preserved so the traffic is correctly processed by the MAC. The ATM CS protocol data unit (PDU) consists of an ATM CS PDU header and an ATM CS PDU payload, which is the same as the payload of the ATM cell. Payload header suppression (PHS) can be applied to ATM connections in order to save transmission resources.

The packet CS allows to interconnect the 802.16 system to systems based on packet protocols such as IP, Point-to-Point Protocol (PPP), or Ethernet. Higher layers service data units (SDU) originating from a bridge or router are mapped onto separate connections identified by a connection identifier (CID) according to criteria such as for example, destination address or priority. Also here, PHS can be optionally applied.

4.4 802.16 SYSTEM PROFILES

The 802.16-2004 standard allows for a multitude of options to compose the air interface of a BWA system. Therefore, the notion of system profiles has been developed. System profiles specify MAC and PHY features that are most typical for implementation. All the mandatory features from the standard remain mandatory in the system profiles. Optional features from the standard may appear in a profile

Table 4.4

MAC and PHY Profiles for WirelessMAN-OFDMA and WirlessHUMAN (-OFDMA). (*From:* IEEE Std. 802.16-2004 [6].)

Id	MAC/PHY	Key feature
OFDMA_profM1	MAC	PMP
OFDMA_profP1	PHY	1.25-MHz channel, licensed
OFDMA_profP2	PHY	3.5-MHz channel, licensed
OFDMA_profP3	PHY	7-MHz channel, licensed
OFDMA_profP4	PHY	14-MHz channel, licensed
OFDMA_profP5	PHY	28-MHz channel, licensed
OFDMA_profP6	PHY	10-MHz channel, unlicensed
OFDMA_profP7	PHY	20-MHz channel, unlicensed

as required or conditionally required. Thanks to such transparent classification of multiple options, equipment vendors can build truly interpretable systems.

Each PHY layer of the 802.16-2004 standard has its own set of system profiles. For WirelessMAN-SC, system profiles consist of MAC and PHY profiles. For WirelessMAN-SCa, WirelessMAN-OFDM, WirelessMAN-OFDMA and WirelessHUMAN, the system profiles include MAC, PHY, and RF (specifying different channelization options) profiles, as well as duplexing specification and transmit power class. Because this book is primarily focused on OFDM/OFDMA, in Tables 4.3 and 4.4 we outline PHY and MAC system profiles for WirelessMAN-OFDM and WirelessMAN-OFDMA air interfaces. It should be noted that for MAC the key feature differentiating the system profiles is the applied network topology (PMP or mesh), while for PHY, this is the channel bandwidth. To fully specify a given system profile, it is also required to choose duplexing method, RF profile, and power class. Full discussion of system profiling exceeds the scope of this book and the reader is referred to the 802.16-2004 standard [6].

4.5 IEEE 802.16 SYSTEM SCENARIO AND PARAMETERS

In the next three chapters of this book we shall describe the state of the art RRM algorithms for OFDMA systems. Since OFDMA is one of the key techniques applied in PHY layer for IEEE 802.16, in this subsection we explain the relations between the IEEE 802.16 standard focusing on WirelessMAN-OFDMA air interface and

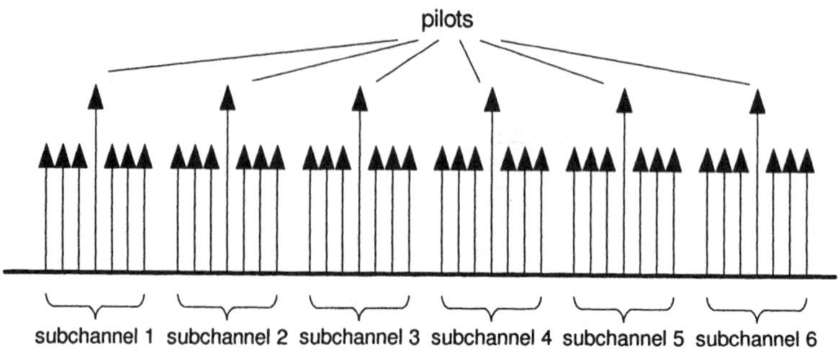

Figure 4.2 Grouping of adjacent subcarriers into OFDMA subchannels.

RRM algorithms treated further in this book. Also, we provide basic system parameters derived from the WirelessMAN-OFDMA scenario that are used further in this book in system analysis and numerical experiments.

Most of the research work on OFDMA assumes per-subcarrier granularity of the frequency resource. Such assumption simplifies the system model. However, in most cases, and in particular in adaptive systems, it requires one to possess perfect knowledge about channel conditions also with per-subcarrier granularity. Obtaining such knowledge would certainly be a difficult task. This is one of the reasons why in WirelessMAN-OFDMA the per-subchannel granularity for resource allocation or system mode adaptation seems to be a more practical choice. Our intention in this book is to both preserve the per-subcarrier reasoning that is present in research literature and be able to apply its conclusions to practical implementations such as the WirelessMAN-OFDMA air interface of the IEEE 802.16 standard. We explain it in the following.

In the WirelessMAN-OFDMA the number of available subcarriers equals N =2,048. It is a relatively high number assuming moderate multipath, especially if we want to simulate RRM algorithms allocating resources with per-subcarrier granularity. Therefore, instead of considering each separate 802.16 subcarrier, we consider a separate subchannel, which is built by grouping a number of adjacent subcarriers. Then, the need for knowledge of radio channel quality is reduced from per-subcarrier to per-subchannel resolution and resources are allocated on per-subchannel basis. Such grouping is allowed as part of the WirelessMAN-OFDMA, where subcarriers constituting a subchannel may but need not be adjacent. Picking nonadjacent subcarriers can be justified in systems where subchannels are randomly assigned to users, for example by the use of frequency hopping, so that such an

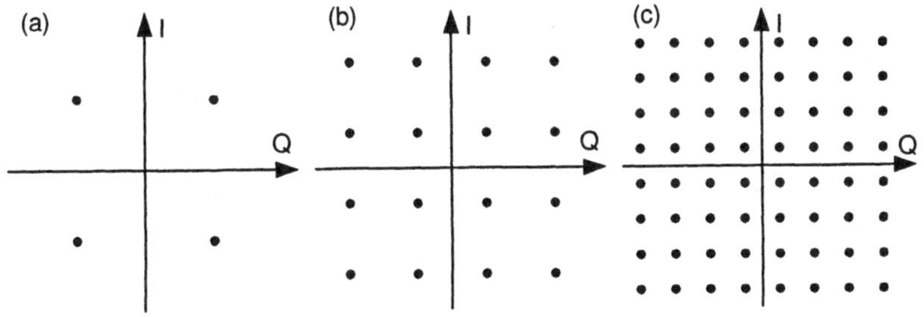

Figure 4.3 M-QAM constellation: (a) 4-QAM (QPSK), (b) 16-QAM, and (c) 64-QAM.

approach provides extra frequency diversity. However, in this book we concentrate on the allocation methods, where resources are assigned adaptively based on the knowledge of the underlying channel conditions. Thus, grouping of the adjacent subcarrier is more plausible here, since fading on adjacent subcarriers is likely to be correlated, especially for moderate multipath. In addition, assuming a flat-fading subchannel, it is sufficient to appoint one subcarrier per subchannel as a pilot to represent the quality of all the component subcarriers, as is shown in Figure 4.2. Then, the required CSI is derived from the pilot measurements. In effect, based on this CSI the RRM algorithm makes decisions about allocation of subchannels to users and setting the transmit power and modulation level per each subchannel. This means that all subcarriers constituting a given subchannel will have the same transmit power and modulation level settings.

In order to preserve the per-subcarrier reasoning and to be able to apply its conclusions into the IEEE 802.16 standard, further in this book we retain the per-subcarrier nomenclature and granularity. However, using a low number of these subcarriers (but sufficient to protect from multipath) in system analysis and simulations, a single subcarrier treated in the following chapters represents a single 802.16 subchannel composed according to the methodology presented above. Thus, all conclusions derived from per-subcarrier models hold for 802.16 WirelessMAN-OFDMA system with per-subchannel granularity. We believe that such an approach allows for application of many other efficient radio resource algorithms that are present in the research literature into practical systems such as the IEEE 802.16 standard. As such, they constitute strong candidate techniques for future standard improvements [3].

In order to further settle the system's analysis and numerical experiments in a BWA scenario of the IEEE 802.16 standard, we have applied values for chosen

Table 4.5

Summary of Basic System and Channel Parameters Used Further in the Book in the Systems' Analysis and Numerical Experiments

α	f_c	σ^2_{shadow}	τ_{rms}	K_{factor}	r	BW	N
4.4	2 GHz	18 dB	264 ns	-3 dB	0.6 km	7 MHz	16

system and channel parameters according to [11–13]. They are summarized in Table 4.5 and applied further in this book unless otherwise stated. The path loss exponent α is determined to be $\alpha = 4.4$ to reflect an intermediate path loss terrain assuming BS antenna height of around 30 m. Attenuation at the reference distance d_0 is set to -39 dB for carrier frequency f_c around 2 GHz. Shadowing variance σ^2_{shadow} equals 18 dB. The values of delay spread $\tau_{rms} = 264$ ns and K-factor $K_{factor} = -3$ dB are taken for the modified Stanford University Interim (SUI) channel model SUI 3, which is applicable to situations with low delay spread, low Doppler, low K-factor, and intermediate path loss. Since we allow for high data rate requirements even at the cell borders, the cellular network is of an urban hot spot type, where cell radius $r = 0.6$ km [12]. It is assumed that the BS is equipped with omnidirectional antenna with gain equal to 11 dBi, while user terminals are equipped with directional antennas pointing at the BS with gain of 19 dBi. The system bandwidth is chosen to be $BW = 7$ MHz, which corresponds to one of the WirelessMAN-OFDMA system profiles, namely OFDMA_ProfP3 [6]. The number of available subcarriers (or subchannels taking the 802.16 nomenclature) in most of the cases is $N = 16$, which is a compromise between accuracy of numerical experiments and simulation time. Possible modulation constellations allowed on each subcarrier are restricted to the three types, namely 4-QAM (QPSK), 16-QAM, and 64-QAM, as shown in Figure 4.3. In addition, the following minor assumptions have been taken:

- No subcarriers are reserved as unused subcarriers (guard or DC);
- Only downlink scenario is considered;
- No mobility (or at most low mobility) is assumed to reflect BWA scenario;
- Framing, channel coding, and ARQ are not considered;
- No frequency hopping is applied. The presented methods are based on deterministic adaptive resource allocation rather than random assignment.

4.6 802.16 STANDARDS SUMMARY

The family of IEEE 802.16 standards addresses well the requirements imposed on future BWA systems as presented in Section 1.1. It offers high data rates, it is able to serve a high number of users, while operating under tough conditions of wireless medium and utilizing available resources efficiently. The issues of mobility and cognitive radio approach are planned to be addressed in future releases of 802.16, namely 802.16e and 802.16h. Other advantages include high system responsiveness, which is particularly important in bursty data applications and ability to use different frequency bands, which significantly increases system application in various countries having different bandwidth allocations. Also, it is possible to operate a system in both licensed and license-exempt frequency bands, which allows for network deployment by both big carriers as well as small (local) service providers. Economies of scale, driven by the involvement of a high number of specialized companies and by standardization and certification efforts, give the promise of affordable broadband access technology.

All these system features lead to a broad range of prospective applications and market segments to name a few [1]:

- Residential broadband Internet access;
- Broadband access for small business;
- E1-level access for large companies;
- Backhaul for WLAN hot-spots or cellular (e.g., GSM) base stations;
- Provision of last mile in rural areas;
- Public systems and services.

Though it may seem that WiMAX is addressing some market niches that are targeted by other systems, (such as 3G), in our opinion it should not be treated as a direct competitor threatening 3G. This is because all these technologies are optimized for a certain purpose, thus they have certain distinct features. The optimization problem that many network operators are (or will soon be) faced with is how to combine a puzzle picture by picking the most effective technologies for certain areas, given the market requirements.

References

[1] http://www.wimaxfoum.org, The WiMAX Forum.

[2] Eklund, C., Marks, R.B., and Stanwood, K.L., "IEEE Standard 802.16: A Technical Overview of the WirelessMAN Air Interface for Broadband Wireless Access," *IEEE Communications Magazine,* June 2002, pp. 98-107.

[3] Ghosh, A., Wolter, D.R., Andrews, J.G., and Chen, R., "Broadband Wireless Access with WiMAX/802.16: Current Performance Benchmarks and Future Potential," *IEEE Communications Magazine,* February 2005, pp. 129-136.

[4] IEEE, *Part 16: Air Interface for Fixed Broadband Wireless Access Systems,* IEEE Std 802.16, 2001.

[5] IEEE, *Part 16: Air Interface for Fixed Broadband Wireless Access Systems-Amendment 2: Medium Access Control Modifications and Additional Physical Layer Specifications for 2-11 GHz,* IEEE Std 802.16a (Amendment to IEEE Std 802.16-2001), 2003.

[6] IEEE, *Part 16: Air Interface for Fixed Broadband Wireless Access Systems,* IEEE Std 802.16-2004, October 2004.

[7] Yaghoobi, H., "Scalable OFDMA Physical Layer in IEEE 802.16 WirelessMAN," *Intel Technology Journal,* vol. 8, issue 3, 2004, pp. 201-212.

[8] IEEE, *Part 11: Wireless LAN Medium Access Control (MAC) and Physical Layer (PHY) Specifications. High-speed Physical Layer in the 5 GHz Band,* IEEE Std 802.11a, 1999.

[9] Koffman, I., and Roman, V., "Broadband Wireless Access Solutions Based on OFDM Access in IEEE 802.16," *IEEE Communications Magazine,* April 2002, pp. 96-103.

[10] Alamouti, S.M., "A Simple Transmit Diversity Technique for Wireless Communications," *IEEE JSAC,* vol. 16, no. 8, October 1998, pp. 1451-1458.

[11] IEEE 802.16 Broadband Wireless Access Working Group, *Channel Models for Fixed Wireless Applications,* June 2003.

[12] Greenstein, L.J., Erceg, V., Yeh, Y.S., and Clark, M.V., "A New Path-Gain/Delay-Spread Propagation Model for Digital Cellular Channels," *IEEE Transactions on Vehicular Technology,* May 1997, volume 46, pp. 477-484.

[13] Erceg, V., Greenstein, L.J., Tjandra, S.Y., Parkoff, S.R., Gupta, A., Kulic, B., Julius, A.A., and Bianchi, R., "An Empirically Based Path Loss Model for Wireless Channels in Suburban Environments," *IEEE JSAC,* July 1999, volume 17, pp. 1205-1211.

Chapter 5

RRM algorithms for OFDMA without spectrum reuse

OFDMA matches well to the multiple user scenario: a subcarrier that is of low quality to one user may be of high quality to another user and can be allocated accordingly. In this chapter, the simple scenario of a single-cell OFDMA system is considered, where a subcarrier (or a group of subcarriers) is allocated exclusively to one user, thus no cochannel interference is created. Assuming a use of adaptive modulation, radio resource allocation here involves assignment of subcarriers to users, bit loading, and power control applied on a per-subcarrier basis. This area of resource allocation for single-cell OFDMA is relatively well explored in the literature. Thus, it lays the basis for the next chapters by discussing basic approaches for resource allocation in OFDMA systems.

The chapter is organized as follows. Section 5.1 presents a literature review of research work on RRM for single-cell OFDMA without spectrum reuse. Section 5.2 introduces a single-cell OFDMA system model. Further in Section 5.2, the QoS-support model is introduced and two most common resource allocation problems are formulated, namely minimization of the total transmit power and maximization of the total data rate. Section 5.3 presents and classifies the most important resource allocation algorithms that can be found in the literature. Based on them, a set of most efficient algorithms is composed, which is later used in numerical experiments presented in Section 5.4. Finally, Section 5.5 concludes this chapter.

5.1 LITERATURE REVIEW

This section provides an overview of the research work done in the area of RRM for OFDMA, focusing on single-cell systems with exclusive assignment of subcarriers to users. The overview is started by presenting early work on OFDMA.

5.1.1 Early work on OFDMA

The application of OFDM to multiuser systems can be dated back to the mid-1990s. Wei proposed in his Ph.D. thesis [1], a multiuser OFDM scheme, where each user is assigned one or several subcarriers, which is essentially an OFDMA idea. In [2], Sari et al. proposed to apply OFDMA for the return channel of a cable digital television. In [3], Rohling and Grunheid compared the performance of OFDMA (the name OFDM-FDMA is used in the paper) with OFDM-TDMA and OFDM-CDMA and suggested adaptive allocation of subcarriers to users, taking into account the information about the radio channel quality. OFDM in combination with FDMA and TDMA is also considered in [4–6] as one of the candidate air interfaces for UMTS proposed by Telia. In this proposal, a user is assigned a number of so-called symbol blocks, where a symbol block is a group of adjacent OFDM subcarriers and adjacent OFDM symbols. A similar approach has also been applied in the WirelessMAN-OFDMA air interface of the IEEE 802.16 standard [7] . The technique of adaptive channel allocation is suggested, where the traffic is placed on the least interfered subcarriers in order to make frequency planning obsolete. Another proposal of an OFDM-based air interface for UMTS was provided by Sony [8] and [9]. In this proposal, a user is allocated so-called "band-slots," which are blocks of 24 adjacent subcarriers. In addition to this subcarriers' grouping, within a 4-slot TDMA frame frequency hopping is applied to hop over different band-slots. Therefore, this scheme falls into the category of interference-averaging techniques. Yet another proposal for the UMTS air interface was to apply a combination of OFDM and TDMA [10]. Earlier, OFDM-TDMA was studied by Rohling and Grunheid in [3, 11, 12]. Also in these papers, simple schemes are proposed for improving performance of such systems by leaving out weak subcarriers and/or applying different modulation modes on each subcarrier.

In all of these early works, OFDM was combined in a straightforward manner with other contemporary multiple access methods, such as TDMA or FDMA. Resource allocation has not yet been treated as an optimization problem. Also, the QoS awareness was weak. Interference was at best averaged or avoided. However, these ideas and research results provided incentives for further studies indicating, among other things, the potential of exploiting multiuser diversity by adaptive allocation of subcarriers to users based on the knowledge of channel conditions.

5.1.2 OFDMA with exclusive allocation of subcarriers to users

The idea of combining adaptive modulation with adaptive subcarrier allocation was introduced by Wong et al. in [13]. Adaptive subcarrier allocation allowed to efficiently distribute subcarriers by exploiting the phenomenon, where a subcarrier, which is of low quality to one user, is not likely to be of low quality to all other users, since users' fading patterns are uncorrelated. Also, this work is the first where resource allocation for OFDMA is treated as an optimization problem. Wong's optimization objective was to minimize the sum transmit power subject to individual user's rate requirements. The original combinatorial optimization problem was exchanged for another, similar problem, where it was allowed that the number of bits allocated to users could be a real value (and not only integer). The proposed solution based on the Lagrangian method is highly complex and consists of two phases: (a) allocation of subcarriers and transmit powers to users, and (b) bit loading, which follows single-user water-filling. In [14], Wong et al. presented a fairly simple solution of greedy distribution of subcarriers to users assuming fixed modulation. The initial greedy assignment is followed by iterative subcarrier swapping between those users for which it leads to a total power reduction.

The paper by Rhee et al. [15] on adaptive subcarrier allocation considered another optimization problem, that is maximization of the minimal capacity of all users. In addition to the optimal solution, a simple suboptimal solution was proposed, which is based on sequential allocation of subcarriers with best gains to users. One of the assumptions then was to use identical power for all subcarriers, since it is claimed that a "flat transmit PSD would hardly reduce the data throughput of a multiuser OFDM system" [15].

In [16], Yin et al. introduced the idea of splitting the allocation process into two phases: (a) resource allocation phase that is determination of the number of subcarriers and amount of power required for each user, and (b) the allocation of specific subcarriers to specific users. This idea was later adopted by many researchers. The two allocation phases are followed by bit loading, given the previous allocations. The considered optimization objective is to maximize the sum rate subject to individual rate requirements and a total power mask (maximum available transmit power).

The same approach of splitting the allocation process into two phases was applied by Kivanc et al. in [17]. However, here the allocation problem is to minimize the total power subject to users' individual rate requirements. The scheme for the determination of the number of required subcarriers per user is very similar to the one presented in [16]. For subcarriers-to-users distribution the authors in [17] proposed two methods, both greedy and based on the allocation of best subcarriers to users. Their major advantage is low complexity. The same results are also published in a broader form in [18].

The original complex problems of [13] and [15] were reformulated by Kim et al. in [19] to simpler problems solvable by linear programming. In addition, fairly simple suboptimal algorithms were proposed, where subcarrier allocation is based on the assumption of a fixed and identical modulation scheme for every subcarrier and allocation of the same number of subcarriers for each user. This was followed by a greedy loading of bits on subcarriers requiring least transmit power. It was shown that both optimal and suboptimal solutions outperform the solutions proposed in [13, 15].

The idea of a two-phase suboptimal algorithm from [16] and [17] was followed by Pfletschinger et al. in [20]. The considered allocation problem is to minimize the total transmit power subject to individual rate requirements and individual power constraints since the considered scenario is the uplink (although the proposed algorithm with slight modifications also applies to the downlink). The part of the algorithm responsible for the allocation of subcarriers to users is much simpler than in [16] and is actually very similar to [14]. The additional feature is that users are allocated subcarriers in the order determined by the users' priorities. The two-phase algorithm is followed by a per-subcarrier bit and power allocation.

In [21], Pietrzyk and Janssen considered a problem of subcarriers' distribution among users, assuming fixed modulation order per each subcarrier. The solution is similar to the one in [14], but it allows for including users' individual BER requirements in the allocation process. This leads to power savings, when compared to the worst-case design. Moreover, the considered resource allocation algorithm is fit into a general contract-based QoS / CoS (cost of service) framework, which allows for more efficient management of resources on the system level.

This group of research work on OFDMA with exclusive allocation of subcarriers to a user is relatively well explored in the literature. Many different optimization problems are considered and various algorithms are proposed. Therefore, this area is a good source for basic solutions for subcarrier allocation and bit loading, which, after modifications, can be applied into more complex system scenarios such as those where cochannel interference is present. Such systems are addressed in later chapters of this book.

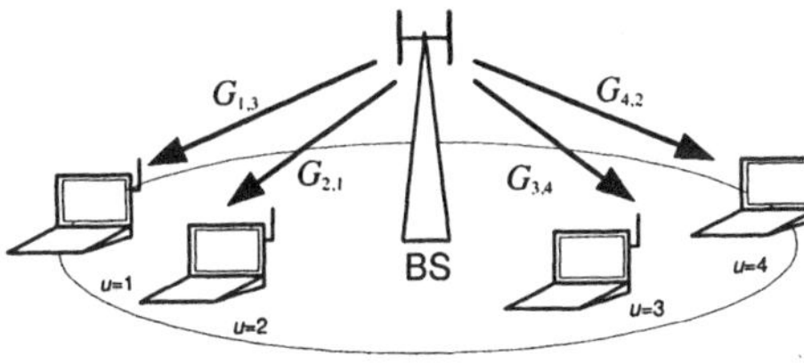

Figure 5.1 One-cell downlink OFDMA system. Example for $U = 4$ users, each allocated one subcarrier exclusively. The channel gains of subcarrier $n, n = 1, \ldots, N$ between BS and user $u, u = 1, \ldots, U$ is indicated by $\boldsymbol{G} = [G_{u,n}]$.

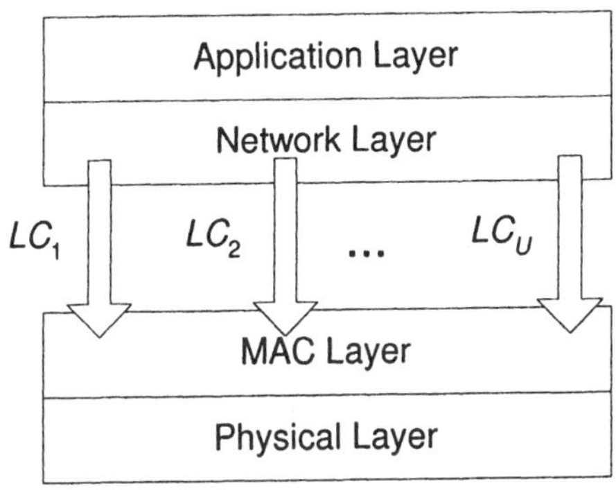

Figure 5.2 System layer model with logical channels (LC) for each user.

5.2 SYSTEM MODEL AND PROBLEM FORMULATION

In this section, a single-cell downlink OFDMA system model, together with an appropriate QoS model, is presented. Also, we explain ways of application of the presented system model to standard systems such as those based on IEEE 802.16 [7]. This is followed by a formulation of the radio resource allocation problems for the two most common system cases as viewed from the perspective of traffic handling, namely constant bit rate (CBR) and variable bit rate (VBR) systems.

5.2.1 System model

The system block model is depicted in Figure 5.1. We consider one cell downlink communication from a single base station (BS) to U user terminals, also referred to

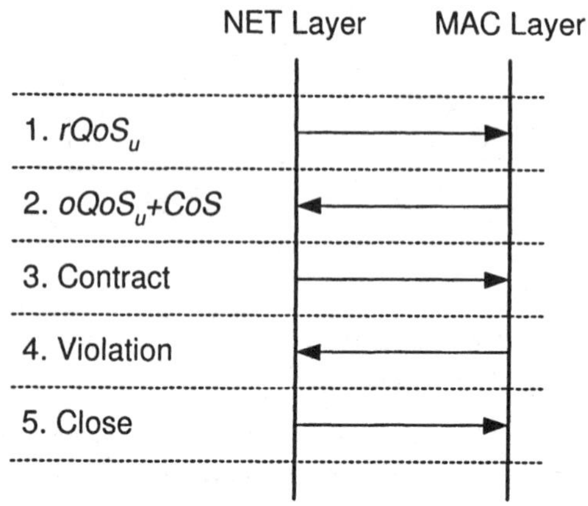

Figure 5.3 Communication between system layers.

as subscriber stations (SS), over a total bandwidth BW, which is partitioned into N OFDM subcarriers, $BW = N\Delta f$. The width Δf of each subcarrier should be smaller than the coherence bandwidth in order to overcome frequency selective fading. Then, instead of frequency selective channel BW we have N frequency nonselective, thus flat fading, subcarriers. Channel gain of each subcarrier n, as perceived by user u is expressed by $\boldsymbol{G} = [G_{u,n}]$, which includes the effects of flat fading, path loss, and shadowing. Signals transmitted on subcarriers undergo AWGN with zero-mean and variance σ^2. $\boldsymbol{G}$ and σ fully describe the channel and they are referred to as channel state information (CSI). It is assumed that the BS has a perfect knowledge of CSI. This knowledge can be obtained by using pilots or applying TDD, where, assuming channel reciprocity, CSI for the downlink transmission can be obtained from the uplink transmission. The user equipment is fixed (no mobility) so that the channel is slowly varying and the Doppler spread can be neglected. This, together with the assumption of perfect time and frequency synchronization, gives a system free from inter-channel interference (ICI). Users are assumed to be uniformly distributed over the cell area. BS is equipped with omnidirectional antenna, while user terminals are equipped with directional antennas pointing at the BS. The maximal BS transmit power is denoted by P_{max}.

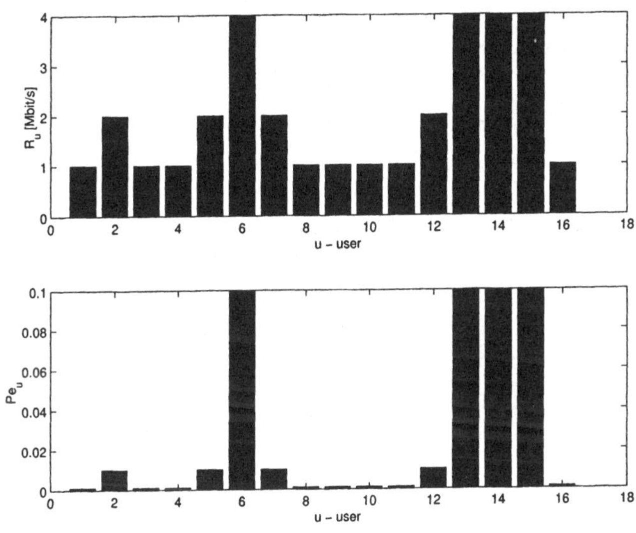

Figure 5.4 Example realization of traffic profile for $U = 16$ users and three services. *Service 1*: $R_u = 1$ Mbit/s, $Pe_u = 10^{-3}$ for 50% of all users, *Service 2*: $R_u = 2$ Mbit/s, $Pe_u = 10^{-2}$ for 25% of all users, and *Service 3*: $R_u = 4$ Mbit/s, $Pe_u = 10^{-1}$ for 25% of all users.

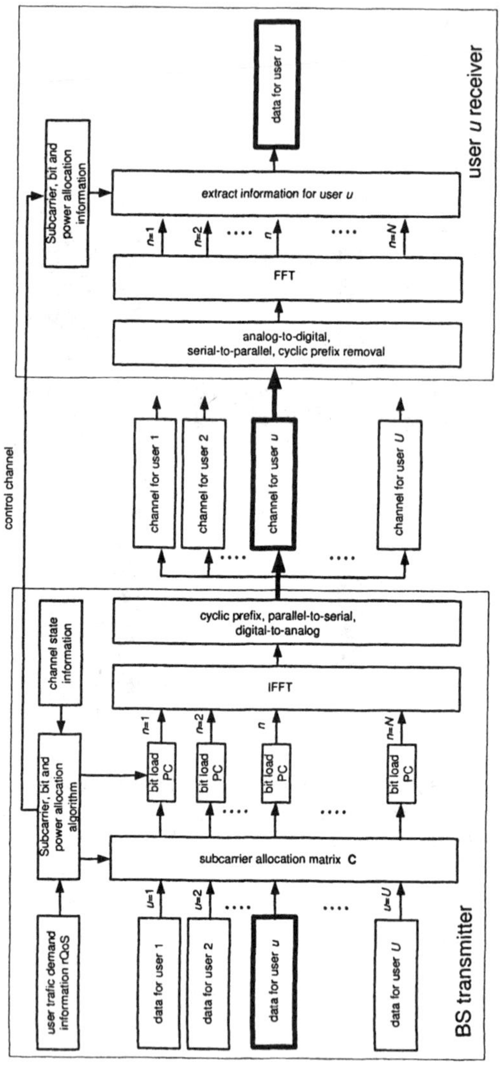

Figure 5.5 Block model of a single-cell OFDMA system with exclusive assignment of subcarriers to users.

For simplicity we assume that the system consists of four communication layers, namely application, network (NET), medium access control (MAC), and physical (PHY), as shown in Figure 5.2. When designing a QoS-aware system it is important to note that the user-perceived QoS characteristics that are visible at the application layer are usually subjective "soft" measures that should be translated into the objective measurable QoS parameters understood by the lower layers [22]. The NET and the MAC layers communicate with each other based on the idea of contracts [23]. This is depicted in Figure 5.3, where for each new user u, the NET layer is requesting a certain QoS profile by indicating required bit rate and bit error rate (BER), $rQoS_u = [R_u, Pe_u]$. Other parameters may also be considered, but these two have been chosen as the most representative for the underlying PHY layer. The MAC estimates if such $rQoS_u$ is feasible and responds to NET layer indicating QoS profile that can be offered (supported), $oQoS_u = [oR_u, oPe_u]$ and the cost that has to be paid CoS - cost of service. This cost can be for example the required transmit power. The communication link over the air interface is established if the offer fulfills the requirements at the acceptable cost[1]. Otherwise, a new user is blocked. Another option might be to gracefully degrade the QoS of other users. However, in this book we do not apply this strategy aiming at provision and maintenance of QoS of already established links. The presented QoS model is clearly a connection-oriented one. Example realizations of traffic profiles for three services are shown in Figure 5.4.

The block model of the considered downlink OFDMA system is depicted in Figure 5.5. At the BS, users are assigned subcarriers according to the allocation matrix $\boldsymbol{C} = [C_{u,n}]$, indicating that a subcarrier n is allocated: $C_{u,n} = 1$ or is not allocated: $C_{u,n} = 0$, to user u. In the considered system, a subcarrier is assigned exclusively to a single user, which means that $\sum_{u=1}^{U} C_{u,n} = \{0, 1\}; n = 1, \ldots, N$. After allocation, each assigned subcarrier n is loaded with $b_{u,n}$ bits according to the bit loading matrix $\boldsymbol{B} = [b_{u,n}]$, $b_{u,n} = 2, 4, \ldots, b_{max}, n = 1, \ldots, N$. This is realized by choosing appropriate square $M_{u,n}$-QAM modulation levels, where $b_{u,n} = log_2(M_{u,n})$. The resulting signal at each subcarrier is power controlled according to the power allocation matrix $\boldsymbol{P} = [P_{u,n}]$, with $\sum_{u=1}^{U} \sum_{n=1}^{N} P_{u,n} \leq P_{max}$. Subsequently, the set of N parallel data streams is processed by the IFFT, a cyclic prefix (CP) is added and the symbols are converted into a serial stream. Finally, the digital to analog conversion is applied and the resulting OFDM signal is transmitted. At the receiver of a particular user u, the signal influenced by the channel $\boldsymbol{G}$ and AWGN noise is received. After inverting the transmitter operations,

1 This model is based on discussions in the UbiCom project at the Delft University of Technology.

we extract the appropriate data addressed to user u, based on the allocation matrix $\boldsymbol{C}$ communicated to all users through a separate control channel.

The average power $P_{u,n}$ of a square QAM constellation at subcarrier n allocated to user u, can be expressed by [24]

$$P_{u,n} = \frac{d_{u,n}^2(M_{u,n} - 1)}{6}, \tag{5.1}$$

where $d_{u,n}$ is the distance between the constellation points. The probability of symbol error $Pe_{u,n}$ in QAM can be closely approximated by [24]

$$Pe_{u,n} = 4 \cdot Q\left(\frac{G_{u,n}d_{u,n}}{2\sigma}\right), \tag{5.2}$$

where σ^2 is the AWGN power.

By combining (5.1) and (5.2), the following formula can be obtained, determining the transmit power $P_{u,n}$ required to carry $b_{u,n} = log_2(M_{u,n})$ bits by user u on subcarrier n with the assumed error probability Pe_u under channel conditions described by $\boldsymbol{G}$ and σ

$$P_{u,n} = \frac{2^{b_{u,n}} - 1}{T_{u,n}}, \tag{5.3}$$

where $T_{u,n}$ is the normalized channel gain, defined as

$$T_{u,n} = \frac{G_{u,n}^2}{\sigma^2 \Gamma_{u,n}}, \tag{5.4}$$

where Γ is often referred to as the SNR gap, defined as [24]

$$\Gamma_{u,n} = \frac{1}{3}\left[Q^{-1}\left(\frac{Pe_u}{4}\right)\right]^2. \tag{5.5}$$

The Q-function in the above formula is defined as [25]

$$Q(x) = \frac{1}{\sqrt{2\pi}} \int_x^\infty e^{-t^2/2} dt. \tag{5.6}$$

Given the transmit power $P_{u,n}$, the bit loading $b_{u,n}$ that can be achieved on subcarrier n allocated to user u under channel conditions described by $G_{u,n}$ and σ

is determined by

$$b_{u,n} = \log_2 \left(1 + \frac{\gamma_{u,n}}{\Gamma_{u,n}} \right), \tag{5.7}$$

where the signal-to-noise ratio is defined as

$$\gamma_{u,n} = \frac{P_{u,n} G_{u,n}^2}{2\sigma^2}. \tag{5.8}$$

5.2.2 Resource allocation problems

In the following, the typical resource allocation problems are presented for two downlink OFDMA systems: constant bit rate (CBR) system and variable bit rate (VBR) system.

5.2.2.1 Constant Bit Rate system

In the CBR system the aim is to achieve a deterministic, fixed bit rate oR_u for each user $u = 1, \ldots, U$, while at the same time minimizing the total transmit power. The CBR allocation problem can be stated as follows

$$\min \sum_{u=1}^{U} \sum_{n=1}^{N} C_{u,n} P_{u,n} \tag{5.9}$$

subject to

$$R_u \leq oR_u = \sum_{n=1}^{N} C_{u,n} b_{u,n}; \; u = 1, \ldots, U, \tag{5.10}$$

$$P_{max} \geq P_{total} = \sum_{u=1}^{U} \sum_{n=1}^{N} C_{u,n} P_{u,n}, \tag{5.11}$$

$$\sum_{u=1}^{U} C_{u,n} \leq 1; \; n = 1, \ldots, N, \tag{5.12}$$

$$C_{u,n} \in \{0, 1\}, \; u = 1, \ldots, U; \; n = 1, \ldots, N. \tag{5.13}$$

Constraint (5.10) expresses the user data rate requirement R_u versus offered data rate oR_u, constraint (5.11) is a limitation of a resulting total transmit power P_{total}, and constraint (5.12) indicates that a subcarrier can be allocated to at most one user within a cell.

5.2.2.2 Variable Bit Rate system

In the VBR system with rate guarantees, which is dual to the CBR problem, the aim is to maximize the sum data rate of the system using all the available power under the condition of satisfying the minimal user data rate requirements. The VBR allocation problem can be stated as follows

$$\max \sum_{u=1}^{U} \sum_{n=1}^{N} C_{u,n} b_{u,n} \tag{5.14}$$

subject to

$$R_u \leq oR_u = \sum_{n=1}^{N} C_{u,n} b_{u,n}; \; u = 1, \ldots, U, \tag{5.15}$$

$$P_{max} \geq P_{total} = \sum_{u=1}^{U} \sum_{n=1}^{N} C_{u,n} P_{u,n}, \tag{5.16}$$

$$\sum_{u=1}^{U} C_{u,n} \leq 1; \; n = 1, \ldots, N, \tag{5.17}$$

$$C_{u,n} \in \{0, 1\}, \; u = 1, \ldots, U; \; n = 1, \ldots, N. \tag{5.18}$$

The meaning of the constraints is the same as in the CBR problem (5.9)-(5.13).

5.2.2.3 Modification for the uplink

In order to apply problems (5.9)-(5.13) and (5.14)-(5.18) to the uplink scenario, constraints (5.11) and (5.16) have to be exchanged for the following relation

$$P_u^{max} \geq P_u^{total} = \sum_{n=1}^{N} C_{u,n} P_{u,n}; \; u = 1, \ldots, U. \tag{5.19}$$

This new constraint limits the individual transmit power level of each user.

5.3 POSSIBLE SOLUTIONS TO THE RRM PROBLEMS

5.3.1 Review of algorithms from the literature

This subsection classifies the algorithms for radio resource allocation for single-cell OFDMA systems with exclusive assignment of subcarriers to users, which were mentioned in Section 5.1. The algorithms can be split into: (a) those aiming at optimal solutions, which tend to be very computationally intensive [13, 15, 19], and (b) those, which are less complex, offering good performance but are not optimal [14–17, 19, 20, 26, 27]. Low complex algorithms can be further split into: (i) those solving the CBR problem [14, 17–20, 27], and (ii) those solving the VBR problem [15, 16, 26]. In some papers, more than one algorithm is proposed. In this subsection and further in the book the focus is on low-complex algorithms for the CBR downlink OFDMA scenario as the most representative for deterministic connection-oriented downlink BWA scenario, such as WiMAX.

Most low-complex algorithms for a CBR system are based on the approach of decoupling the allocation process into the following radio resource allocation phases:

- *Initial resource allocation phase*, where the number of subcarriers and/or amount of transmit power to be assigned to each user is determined,
- *Subcarriers-to-users allocation phase*, where assignment of particular subcarriers to particular users (most often based on choosing the best channel gains) is performed,
- *Per-subcarrier bit loading and power control phase*, where the algorithm determines, per each allocated subcarrier, the number of bits to be carried (i.e., the modulation level) and the corresponding transmit power.

This split was initially proposed in [16] and [17] and was followed by most of the other authors. This is motivated by the following observation taken from [16]: "... it is plausible to assume that the resources (number of subcarriers, power) allocated to a particular user depends primarily on its rate requirement R_u and its average channel/noise conditions. On the other hand, the optimality of subcarrier assignment and bit loading is clearly more sensitive to the local channel/noise characteristics."

In the following, instead of discussing each complete algorithm separately, the above-presented three-phase decoupling is applied and for each phase possible allocation approaches that can be found in the literature and research work are discussed.

5.3.1.1 Initial resource allocation phase

In this phase, the amount of resources to be allocated to each user is determined. For CBR systems, this concerns determination of the number of subcarriers for each user, $s_u, u = 1, \ldots, U$. The following options for the determination of s_u can be identified:

- *Option 1*

  Assume that the modulation order M is fixed for all subcarriers. Then, the number of subcarriers is proportional to the user's rate requirement and can be expressed by [14,21]

$$s_u = \lceil T_e R_u / \log_2 M \rceil; \quad u = 1, \ldots, U, \tag{5.20}$$

  where T_e is the effective OFDM symbol duration. The advantage of this approach is its simplicity. However, a user that is far from the BS should rather satisfy its requirements by allocating more subcarriers than by increasing the modulation level. This scheme does not allow for it since the same number of subcarriers is allocated to each user.

- *Option 2*

  In this option the algorithm initially assigns to each user the least number of subcarriers that would satisfy a user's rate requirement. Assuming that each subcarrier can be loaded to at most b_{max} bits per symbol, the minimal number of subcarriers can be expressed by

$$s_u = \lceil T_e R_u / b_{max} \rceil; \quad u = 1, \ldots, U. \tag{5.21}$$

  Subsequently, based on the average (over the subcarriers) channel conditions, the algorithm allocates the remaining subcarriers (if any) to users requiring the least total transmit power. User average transmit power is defined as [16,20,27],

$$\overline{P}_u(s_u) = s_u(2^{T_e R_u / s_u} - 1)/\overline{T}_u; \quad u = 1, \ldots, U, \tag{5.22}$$

  where the user average normalized channel gain is defined by

$$\overline{T}_u = \frac{1}{N} \sum_{n=1}^{N} T_{u,n}; \quad u = 1, \ldots, U. \tag{5.23}$$

  The complete routine determining the number of subcarriers per user [17,27], is described in Algorithm 5.1.

Algorithm 5.1: Determination of the number of subcarriers per user for a single-cell CBR OFDMA system with exclusive assignment of subcarriers to users.

---

input : $\overline{P}_u, R_u, T_e, b_{max}, u = 1, \ldots, U$
output: s_u

begin
  $s_u = \lceil T_e R_u / b_{max} \rceil; \;\; u = 1, \ldots, U$
  while $\sum_{u=1}^{U} s_u < N$ **do**
    for $u = 1$ **to** U **do**
      $\Delta P_u = \overline{P}_u(s_u) - \overline{P}_u(s_u + 1)$
      $w = \text{arg max}_u \, \Delta P_u$
      $s_w = s_w + 1$
      $\overline{P}_w = \overline{P}_w + \Delta P_w$
end

Additionally, before Algorithm 5.1, we can use Algorithm 5.2, controlling the demand for data rate [17], in order to ensure that the data rate requirements can be satisfied using N subcarriers.

Algorithm 5.2: Data rate demand control for a single-cell CBR OFDMA system.

---

input : initial $s_u, u = 1, \ldots, U$
output: s_u possible to realize with N subcarriers

begin
  while $\sum_{u=1}^{U} s_u > N$ **do**
    $w = \text{arg min}_u \, s_u$
    $s_w = 0$
end

An interesting derivative of Algorithm 5.1 for the VBR downlink system can be found in [16], while a slightly modified version applicable for the uplink can be found in [20]. In this algorithm, the amount of power assigned to users should be proportional to the number of allocated subcarriers, so that

$$P_a \leq \frac{N_a}{N} P_{max} \tag{5.24}$$

where $P_a = \sum_u P_u$ is the allocated power and $N_a = \sum_u s_u$ is the number of allocated subcarriers. The routine determining s_u for a VBR system is described in Algorithm 5.3.

Algorithm 5.3: Determination of the number of subcarriers per user for a single-cell VBR OFDMA system with exclusive assignment of subcarriers to users.

---

input : $\overline{P}_u, R_u, T_e, P_{max}, u = 1, \ldots, U$
output: s_u

begin
  $s_u = 1; \quad u = 1, \ldots, U$
  while $P_a > \frac{N_a}{N} P_{max}$ **do**
    for $u = 1$ **to** U **do**
      $\Delta P_u = \overline{P}_u(s_u) - \overline{P}_u(s_u + 1)$
      $w = \arg\max_u \Delta P_u$
      $s_w = s_w + 1$
      $\overline{P}_w = \overline{P}_w + \Delta P_w$
      calculate P_a, N_a
end

Algorithm 5.3 stops when $P_a = \frac{N_a}{N} P_{max}$. Defining the power allocated per subcarrier as $P_n = \frac{P_a}{N_a}$ means that the algorithm stops when $P_n = \frac{P_{max}}{N}$. This corresponds to allocating identical power levels for all subcarriers, which is one of the key assumptions exploited in the papers addressing the VBR problem [15,26].

5.3.1.2 Subcarriers-to-users allocation phase

Having determined the number of subcarriers for each user s_u, $u = 1, \ldots, U$ we can assign particular subcarriers to particular users. The following options for this operation may be identified.

- *Option 1*

  Users, one after another, pick the subcarrier with highest normalized channel gain $T_{u,n}$ [14, 27]. The order in which this is done may be controlled by priorities as suggested in [20], however, we do not consider them here. The corresponding routine is described in Algorithm 5.4.

Algorithm 5.4: Initial allocation of subcarriers to users in a single-cell OFDMA system with exclusive assignment of subcarriers to users.

input : $s_u, [T_{u,n}], u = 1, \ldots, U; n = 1, \ldots, N$
output: $[C_{u,n}]$

begin
  $C_{u,n} = 0; \quad u = 1, \ldots, U; \quad n = 1, \ldots, N$
  while $\sum_u s_u > 0$ **do**
    for $u = 1$ **to** U **do**
      $m = \arg\max_{n \in \{\text{not assigned}\}} T_{u,n}$
      $C_{u,m} = 1$
      $s_u = s_u - 1$
end

Algorithm 5.4 is a practical scheme, since it does not require apriori knowledge of particular bit loading to create the cost or profit function (in the case of CBR systems the cost function is the transmit power). This is based on the assumption that the better the chosen channel, the lower the transmit power can be expected. Therefore, bit loading can be performed after allocation of subcarriers to users.

- *Option 2*

  The allocation performed by the use of Algorithm 5.4 can be treated as an initial allocation, since it usually results in an overall cost (transmit power) higher than optimal. It may be further improved by the use of iterative swapping of subcarriers between those users, for which it leads to total cost reduction [14,21]. Let us assume that initial allocation of subcarriers to users is indicated by $[C_{u,n}]$ and minimal power allocated to subcarriers by $[P_{u,n}^{min}]$. The routine iteratively improving initial allocation is described in Algorithm 5.5 and is based on [14,21].

  Another swapping routine, where the aim is to maximize the sum rate instead of minimizing the total power, can be found in [17].

Algorithm 5.5: Swapping of subcarriers between users in a single-cell OFDMA system with exclusive assignment of subcarriers to users.

input : $[P_{u,n}]$, initial $[C_{u,n}]$
output: improved $[C_{u,n}]$ and $[P_{u,n}^{min}]$

begin
  repeat
    for *each pair of users* $(i, j) \in \{1, \ldots, U\}$ **do**
      calculate power reduction vector: $\delta p_{i,j}(n) = P_{i,n} - P_{j,n}$
      $P_{i,j} = \max(\delta p_{i,j})$
      $n_{i,j} = \arg\max_n(\delta p_{i,j})$
      calculate power reduction vector: $\delta p_{j,i}(n) = P_{j,n} - P_{i,n}$
      $P_{j,i} = \max(\delta p_{j,i})$
      $n_{j,i} = \arg\max_n(\delta p_{j,i})$
      calculate $\Delta P = P_{i,j} + P_{j,i}$
    $(\bar{i}, \bar{j}, \bar{n}_{i,j}, \bar{n}_{j,i}) = \arg\max_{u,n} \Delta P$
    if $\max \Delta P > 0$ **then**
      {apply swapping in $[C_{u,n}]$: }
      $(\bar{n}_{i,j}) \rightarrow \bar{j}$
      $(\bar{n}_{j,i}) \rightarrow \bar{i}$
    update power allocation $[P_{u,n}^{min}]$
  until $\Delta P > 0$
end

- *Option 3*

  The problem of assigning subcarriers to users, given the number of subcarriers per each user s_u, is a classical assignment (or matching) problem for which the optimal solution can be found with the use of the so-called Hungarian algorithm [28], which is described in the appendix at the end of the book. In the case of a matching problem a simple reformulation is required as explained in Section 6.3.2 (based on [29]), transforming the matching problem to the equivalent assignment problem, which can then be solved by the Hungarian algorithm [28].

5.3.1.3 Per-subcarrier bit loading and power control phase

In order to benefit from adaptive modulation [30] as the last phase of the resource allocation process, bit loading and power control are performed on a per-subcarrier basis. This phase is very similar to the water-filling known from single-user OFDM systems. The aim for the CBR system is to spend as little transmit power as possible to meet users' rate requirement, while the aim for the VBR system is to spend all the available transmit power aiming at maximization of the system's sum rate provided users' individual rate requirements are satisfied. Two options for per-subcarrier bit loading and power assignment were found in the literature.

- *Option 1*

  The allocation consists of two steps: *initial bit loading* and *bit-load swapping*. The approach was presented in the OFDMA context in [27] but was originally proposed in [31]. In the *initial bit loading* step bits are distributed over subcarriers proportional to the users' rate requirements

  $$b_{u,n} = \lfloor T_e R_u / s_u \rfloor; \quad u = 1, \ldots, U. \tag{5.25}$$

  Then, in the *bit-load swapping* step bits per user are swapped between those subcarriers for which it leads to overall cost reduction according to the routine given in Algorithm 5.6.

- *Option 2*

  This option for bit loading distributes bits to satisfy R_u by loading one bit at a time on the subcarrier that requires least cost (transmit power) [19]. The routine is described in Algorithm 5.7.

Usually in a practical system the number of bits per symbol is restricted and related to the available modulation constellations. Thus, both Algorithms 5.6 and 5.7 should be amended with appropriate control of the available number of bits per symbol. This is omitted here for clarity of explanation.

Algorithm 5.6: Bit load swapping routine for a single-cell OFDMA system with exclusive assignment of subcarriers to users.

input : $[P_{u,n}]$, initial $[b_{u,n}]$, $s_u, u = 1, \ldots, U$
output: improved $[b_{u,n}]$ and $[P_{u,n}^{min}]$

begin
  for $u = 1$ **to** U **do**
    $\Delta P_{u,n}^{+} = (2^{b_{u,n}+1} - 1)/T_{u,n} - (2^{b_{u,n}} - 1)/T_{u,n};$
    $(u,n) \in \{(u,n)|C_{u,n} = 1\}$
    $\Delta P_{u,n}^{-} = (2^{b_{u,n}} - 1)/T_{u,n} - (2^{b_{u,n}-1} - 1)/T_{u,n};$
    $(u,n) \in \{(u,n)|C_{u,n} = 1\}$
    {subcarrier where we prospectively add a bit : }
    $n_1 = \arg\max_n \Delta P_{u,n}^{+}$
    {subcarrier where we prospectively remove a bit : }
    $n_2 = \arg\min_n \Delta P_{u,n}^{-}$
  if $\Delta P_{u,n}^{+} < \Delta P_{u,n}^{-}$ **then**
    {swap a bit between n_1 and n_2: }
    $b_{u,n_1} = b_{u,n_1} + 1$
    $b_{u,n_2} = b_{u,n_2} - 1$
  update power allocation $[P_{u,n}^{min}]$
end

Algorithm 5.7: Greedy bit loading routine for a single-cell OFDMA system with exclusive assignment of subcarriers to users.

input : $[P_{u,n}]$, $R_u, u = 1, \ldots, U$
output: $[b_{u,n}]$, $[P_{u,n}^{min}]$

begin
  $b_{u,n} = 0; u = 1, \ldots, U; n = 1, \ldots, N$
  for $u = 1$ **to** U **do**
    while $oR_u < R_u$ **do**
      $\Delta P_{u,n} = (2^{b_{u,n}+1} - 1)/T_{u,n} - (2^{b_{u,n}} - 1)/T_{u,n}$
      $m = \arg\min_{n \in \text{user } u} \Delta P_{u,n}$
      $b_{u,m} = b_{u,m} + 1$
      $P_{u,m} = P_{u,m} + \Delta P_{u,m}$
      $oR_u = oR_u + b_{u,m}/T_e$
  update power allocation $[P_{u,n}^{min}]$
end

5.3.2 Algorithm for OFDMA with exclusive subcarrier assignment

In this subsection, a selection of routines proposed in the previous subsection, including all the allocation phases, is presented to form a complete RRM algorithm for a single-cell OFDMA system with exclusive assignment of subcarriers. The choice of the component routines is based on qualitative rather than quantitative analysis. This final algorithm, referred to as *OFDMA exclusive*, is also used in the remainder of this book for the evaluation of the OFDMA performance in a single-cell and as an example of a basic scheme for OFDMA RRM. The OFDMA exclusive algorithm consists of the three allocation phases for which the following schemes have been chosen.

5.3.2.1 Initial resource allocation

In this phase, Algorithm 5.1 (i.e., *Option 2*) is chosen [17, 27]. The reasons are as follows: this routine aims at satisfying users' rate requirements with the least possible number of subcarriers and allocates all the available subcarriers to reduce the overall transmit power.

5.3.2.2 Subcarriers-to-users allocation

In this phase, Algorithm 5.4 (i.e., *Option 1*) [14, 20, 27], is chosen for initial assignment, which is followed by Algorithm 5.5 (i.e., *Option 2*) [14,21], to improve the initial assignment. The motivation behind allocating subcarriers to users based on best channel gains is that it is a low-complex option, which leaves the bit loading for the last phase of the resource allocation process.

5.3.2.3 Per-subcarrier bit loading and power allocation

For this phase, Algorithm 5.7 (i.e., *Option 2*) is chosen [19], since it is more suitable for a system with a finite and small number of possible modulation schemes (here 4-QAM, 16-QAM, and 64-QAM). Algorithm 5.6 (i.e., *Option 1*) has the drawback of introducing errors in the initial assignments, which is significant for a small number of possible modulation states.

5.4 NUMERICAL EXPERIMENTS

In order to evaluate the performance of the RRM algorithm for the downlink OFDMA system with exclusive assignment of subcarriers to users, indicated in the following as OFDMA exclusive, simulations have been performed using the following system parameters as introduced in Chapter 4. In addition, it is assumed that users are uniformly distributed within the cell and each user requires the same data rate $R_u = R, u = 1, \ldots, U$, though the model allows to set R_u per user individually. An example realization of the channel gains showing frequency selectivity of the multipath channel is presented in Figure 5.6.

5.4.1 Example allocation patterns

Example bit loading, subcarrier assignment, and power allocation patterns for the OFDMA exclusive algorithm with adaptive modulation are shown in Figure 5.7. The total required data rate $R_{total} = U \cdot R_u$ equals 28 Mbit/s and is lower than the maximal system capacity $R_{total}^{max} = N \cdot b_{max}/T_e$, which equals to 42 Mbit/s, where T_e is the effective symbol duration. It can be observed that subcarriers and bit loads are allocated to users adaptively, based on experienced channel gains. Each user gets 16 bits within $T_e = 2.2857$ ms, which gives the required 7 Mbit/s per user. Example allocation patterns for the same channel and traffic conditions but for

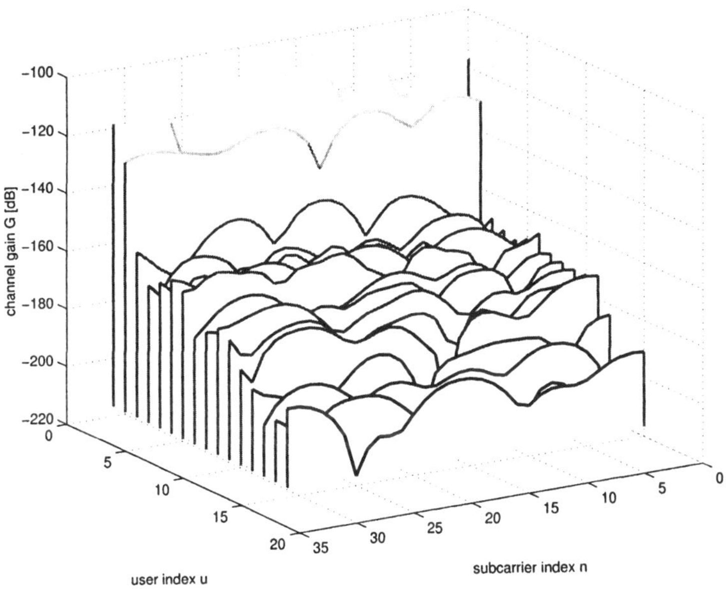

Figure 5.6 Example channel gains $\boldsymbol{G} = [G_{u,n}]$ for $N = 32$ and $U = 16$. User $u = 1$ is the closest to the BS; user $u = 16$ is the furthest. Channel parameters as in the text.

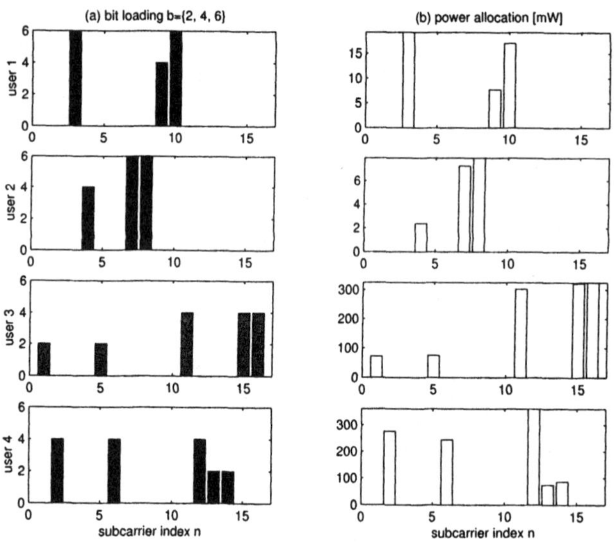

Figure 5.7 Example allocation pattern for the OFDMA exclusive algorithm with adaptive modulation for (a) bits and subcarriers, and (b) transmit power; a single cell, $N = 16, U = 4, R_u = R = 7$ Mbit/s, $Pe_u = Pe = 10^{-4}$, and channel parameters as in the text.

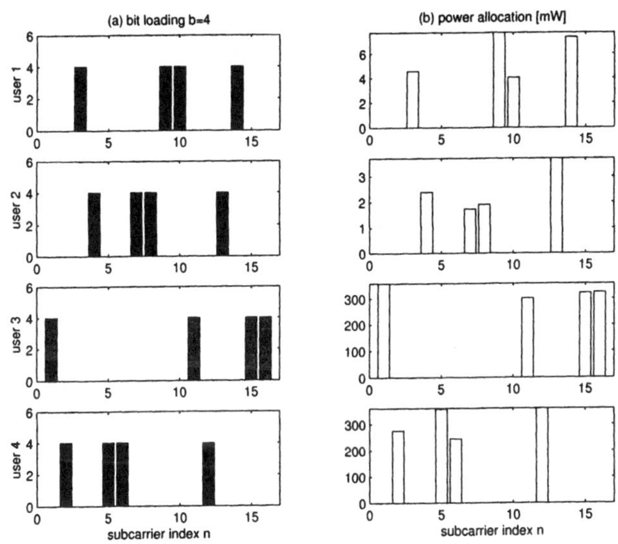

Figure 5.8 Example allocation pattern for the OFDMA exclusive algorithm with fixed modulation (16-QAM with $b = 4$ bits per subcarrier) for (a) bits and subcarriers, and (b) transmit power; a single cell, $N = 16, U = 4$, $R_u = R = 7$ Mbit/s, $Pe_u = Pe = 10^{-4}$, and channel parameters as in the text.

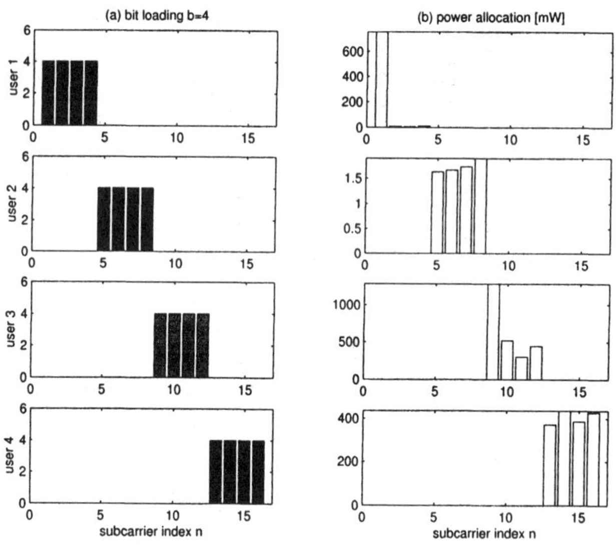

Figure 5.9 Example allocation pattern for FDMA with fixed modulation (16-QAM with $b = 4$ bits per subcarrier) for (a) bits and subcarriers, and (b) transmit power; a single cell, $N = 16$, $U = 4$, $R_u = R = 7$ Mbit/s, $Pe_u = Pe = 10^{-4}$, and channel parameters as in the text.

an OFDMA system employing fixed modulation are depicted in Figure 5.8. Here, only subcarriers are allocated adaptively, since the bit load per subcarrier is fixed to $b = 4$. The system is fully loaded, since the maximal offered total data rate that can be achieved with $N = 16$ subcarriers and $b = 4$ bits per subcarrier within $T_e = 2.2857$ ms equals 28 Mbit/s. The same applies to fixed modulation FDMA with example allocation patterns shown in Figure 5.9. Here, neither modulation levels nor subcarriers are allocated adaptively, which has a detrimental effect on the transmit power, which becomes more explicit in subsequent experiments.

5.4.2 Convergence of iterative swapping

In the following, we validate the convergence of the iterative swapping routine, which is part of the OFDMA exclusive algorithm. Total transmit power defined as $P_{total} = \sum_u \sum_n P_{u,n}^{min}$ has been used as the performance measure. The performance of the OFDMA exclusive scheme converges to the optimal solution obtained by the use of the Hungarian algorithm. For simplicity, a fixed 16-QAM modulation has been applied for all subcarriers. The results averaged over 100 channel / traffic realizations, for $N = 32, U = 8$ and $N = 64$, $U = 16$ are depicted in Figure 5.10 (a) and (b), respectively. It can be observed that the suboptimal iterative swapping algorithm converges well to the optimal reference after several iteration steps. However, it may occasionally happen that for a relatively low number of subcarriers and users (thus low multiuser diversity), the algorithm is trapped in a dead-loop and subsequent iteration steps do not provide any performance improvement. Fortunately, for higher N and U, the algorithm converges closely to the optimal solution. The benefit of the iterative swapping routine is that it allows to scale the allocation performance with time thus allowing to stop its operation when the available time has elapsed.

5.4.3 Influence of individual bit error rate requirements

The next experiment is to evaluate the performance of the OFDMA exclusive algorithm incorporating individual bit error rate requirements Pe_u in the allocation process and compare these results with the ones obtained when a common *worst-case*, most demanding requirement $Pe^{worst} = \min(Pe_u), u = 1, \ldots, U$ is used instead of user-individual bit error rate requirements Pe_u. This approach was applied in [21], in a context of a fixed modulation OFDMA CBR system. Proposal [21], based on Wong's algorithm [14], showed the benefits of differentiation of users' bit error rate requirements and considered explicitly the QoS model for OFDMA

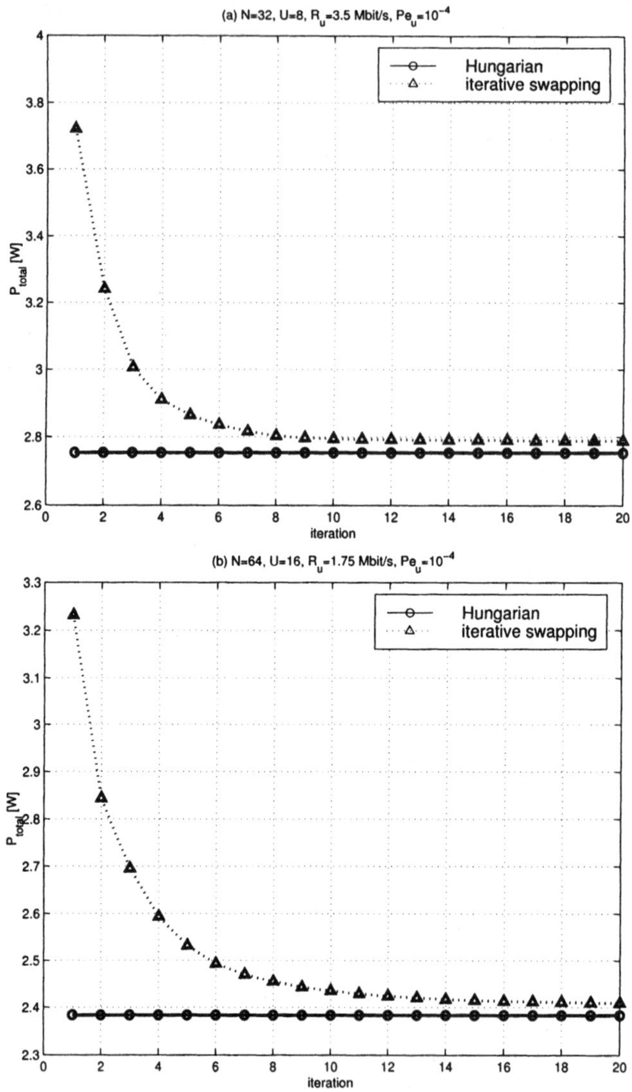

Figure 5.10 Convergence of the iterative swapping routine.

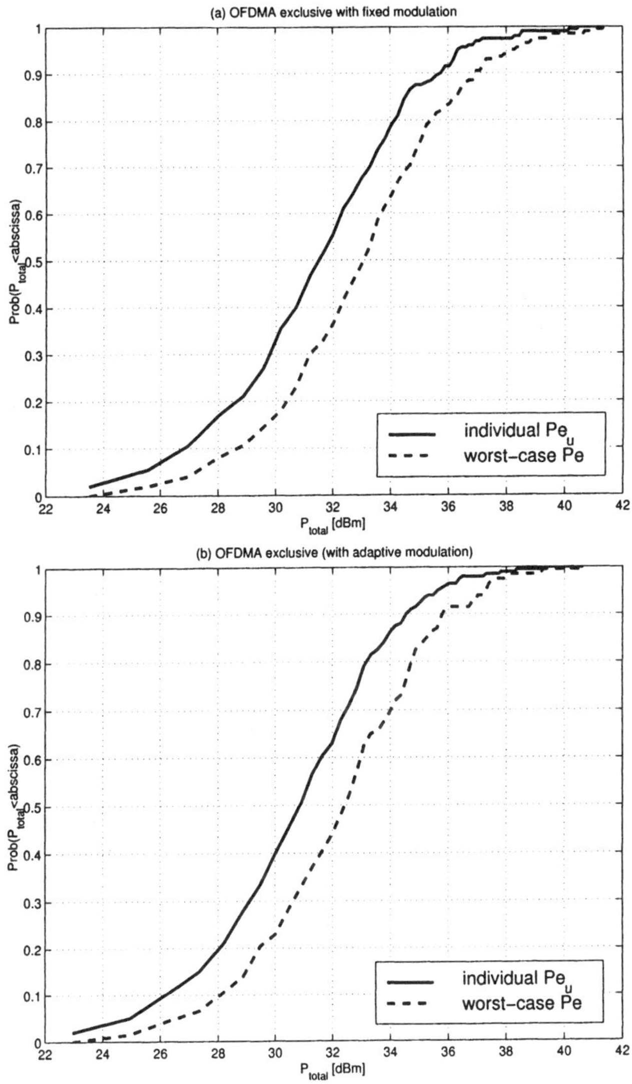

Figure 5.11 Comparison of the cumulative distribution function for the OFDMA exclusive algorithm applying user-individual bit error rate requirements Pe_u and worst-case bit error rate requirement Pe^{worst}. $N = 32$, $U = 8$, $R_u = 3.5$ Mbit/s, 50% of users have $Pe_u(service1) = 10^{-4}$ and the other 50% have $Pe_u(service2) = 10^{-2}$; $Pe^{worst} = 10^{-4}$.

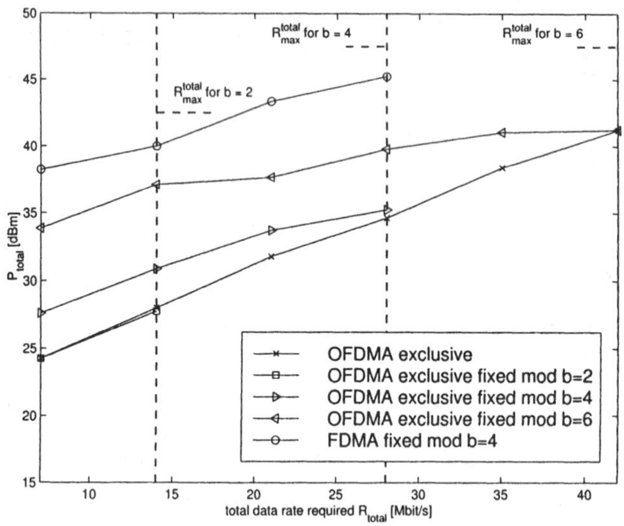

Figure 5.12 Performance comparison of the OFDMA exclusive with adaptive modulation, OFDMA exclusive with various fixed modulations, and FDMA with fixed modulation; $N = 16$, $U = 4$, $Pe_u = Pe = 10^{-4}$.

systems. The results for an OFDMA system applying the OFDMA exclusive algorithm with fixed and with adaptive modulation are depicted in Figure 5.11. The results have been derived from 100 different channel / traffic realizations. It can be observed that a gain of $\sim$ 2 dB is obtained for the example case of two services with $Pe_u(service1) = 10^{-4}$ and $Pe_u(service2) = 10^{-2}$, thus $Pe^{worst} = 10^{-4}$. Though not shown explicitly in this book, higher differences between service error rate requirements lead to even higher performance gains when compared to the worst-case approach. An exception is when all users require the same error rate. Then the performance obtained by application of both approaches (i.e., user-individual and *worst-case*) is the same, since $Pe^{worst} = \min(Pe_u) = Pe_u$, $u = 1, \ldots, U$.

5.4.4 Performance comparison

As the last evaluation, the performance of the following schemes have been compared:

- OFDMA exclusive algorithm with adaptive modulation: 4-QAM, 16-QAM, and 64-QAM;
- Three versions of the OFDMA exclusive algorithm with fixed modulation, each using one of the possible modulations: 4-QAM, 16-QAM, or 64-QAM;
- FDMA with fixed 16-QAM.

The results, derived from 100 different channel and traffic realizations, are shown in Figure 5.12. As a reference the maximal total data rates R_{max}^{total} are indicated, which can be achieved using each of the three modulations. As can be seen in Figure 5.12 the best performance is obtained with the OFDMA exclusive scheme employing adaptive modulation, while the worst performance is obtained with fixed modulation FDMA. It is interesting to note that when the total data rate required equals the maximal system capacity, the performance of each fixed modulation OFDMA exclusive scheme is very close to the performance of the OFDMA exclusive with adaptive modulation. This is because in such situations the OFDMA exclusive scheme with both fixed and adaptive modulation use the same bit loading patterns. In cases when the total traffic required is less than maximal system capacity, the OFDMA exclusive with fixed modulation exhibits worse performance than the OFDMA exclusive with adaptive modulation, since too high (thus too costly) modulation orders for the required traffic are used. The worst performance of fixed modulation FDMA stems mainly from the fact of not using adaptive allocation of subcarriers to users but a fixed assignment, which is done regardless of channel conditions. Another reason is lack of modulation adaptivity.

As a last exercise, for an example total traffic load of half the maximal system capacity (i.e., 21 Mbit/s) of the OFDMA exclusive scheme with adaptive modulation, the cumulative distribution functions are plotted for the considered schemes. The results are shown in Figure 5.13. Again, the best performance is achieved with the OFDMA exclusive scheme using adaptive modulation, with a gain of ~ 8 dB over the worst-case fixed modulation FDMA method.

It can be concluded that the OFDMA exclusive scheme with adaptive modulation is better for systems where the required traffic load cannot be predicted or controlled. For systems where the required traffic load is close to the maximal system capacity (which should in fact be the case of any well-dimensioned system), the performance penalty of the OFDMA exclusive scheme with fixed modulation is negligible, while at the same time the resource allocation algorithm is simpler (since it is unnecessary to perform bit loading). Therefore, using the OFDMA exclusive scheme with fixed modulation is recommended for such systems.

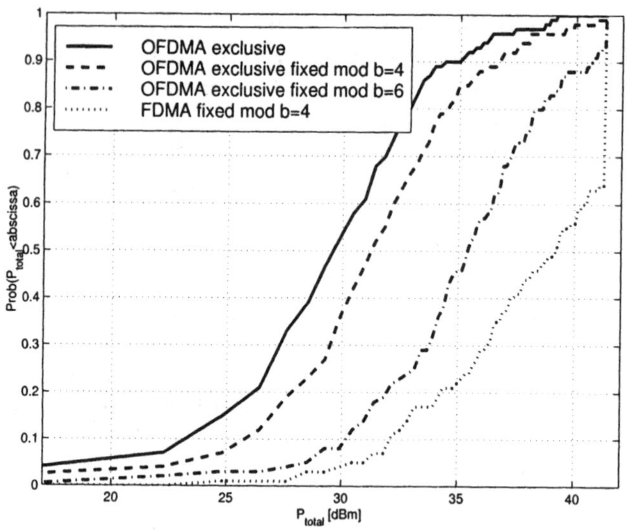

Figure 5.13 Comparison of the cumulative distribution function for OFDMA exclusive with fixed and adaptive modulations, and FDMA scheme with fixed modulation. $N = 16$, $U = 4$, $R_u = R = 5.25$ Mbit/s, which corresponds to the total data rate requirement of 21 Mbit/s; $Pe_u = Pe = 10^{-4}$.

In the remainder of this book, the OFDMA exclusive scheme with adaptive modulation is used as the more general and representative case. In addition, in Chapter 6 more performance evaluations of the OFDMA exclusive scheme with adaptive modulation are provided in comparison with OFDMA RRM allocation methods reusing subcarrier within a cell.

5.5 SUMMARY

This chapter explains some key algorithms from the literature and presents evaluations based on simulations of the chosen options of RRM for a single-cell downlink OFDMA system with exclusive allocation of subcarriers to users. A particular focus has been given to CBR systems where user data rates are deterministic and the system allocation objective is to minimize the total transmit power.

The presented OFDMA system model has been set in the scenario of the IEEE 802.16 standard. Moreover, it is fit into a general contract-based QoS / CoS (cost of service) framework, which allows for more efficient management of resources on the system level.

Simulation results indicate the superiority of adaptive subcarrier allocation over fixed allocation. In addition, subcarrier allocation schemes with and without application of adaptive modulation are compared. It can be concluded that application of fixed modulation is recommended for systems with stable traffic profiles (fixed required total data rate) due to the simpler RRM routine. For all other cases adaptive modulation will result in a better performance.

References

[1] Wei, L., *Single-user and Multi-user Modulation and Detection Methods for Mobile Communications,* University of South Australia, Ph.D. Dissertation, 1994.

[2] Sari, H., Levy, Y., and Karam, G., "Orthogonal Frequency-Division Multiple Access for the Return Channel in CATV Networks," *Proc. ICT,* 1996.

[3] Rohling, H., and Grunheid, R., "Performance Comparison of Different Multiple Access Schemes for the Downlink of an OFDM Communication System," *Proc. IEEE VTC,* pp. 1365-1369, 1997.

[4] van de Beek, et al., "A Conceptual Study of OFDM-based Multiple Access Systems, Part 1: Air Interface Requirements," ETSI Technical Report, 1996.

[5] van de Beek, J.-J., et al., "A Conceptual Study of OFDM-based Multiple Access Systems, Part 2: Channel Estimation in the Uplink," ETSI Technical Report, 1996.

[6] van de Beek, J.-J., et al., "A Conceptual Study of OFDM-based Multiple Access Systems, Part 3: Performance Evaluation of a Coded System," ETSI Technical Report, 1996.

[7] IEEE, *Part 16: Air Interface for Fixed Broadband Wireless Access Systems,* IEEE Std 802.16-2004, October 2004.

[8] Suzuki, M., Bohnke, R., and Sakoda K., "Band Division Multiple Access (BDMA) system. A Novel Approach for Next Generation Mobile Telecommunications System, based on OFDM and SFH-TDMA," *Proc. IEEE VTC,* pp. 1326-1330, 1998.

[9] Suzuki, M., Bohnke, R., and Sakoda, K., "Spectral Efficient Modulation Schemes in a SFH-TDMA Orthogonal Frequency Division Multiplexing (OFDM) Wireless Communication System to support Advanced Services," *Proc. IEEE VTC,* pp. 2202-2205, 1998.

[10] Castro, J.P., Hallmann, E., and Rohling, H., "Downlink OFDM Techniques in 3rd Generation TDMA Based Mobile System," *Proc. IEEE PIMRC,* 1998.

[11] Rohling, H., and Grunheid, R., "Performance of an OFDM-TDMA Mobile Communication System," *Proc. IEEE VTC,* pp. 1589-1593, 1996.

[12] Rohling, H., and Grunheid, R., "Adaptive Coding and Modulation in an OFDM-TDMA Communication System," *Proc. IEEE VTC,* pp. 773-776, May 1998.

[13] Wong, C.Y., et al., "Multiuser OFDM with Adaptive Subcarrier, Bit, and Power Allocation," *IEEE Journal on Selected Areas in Communications,* October 1999, volume 17, pp. 1747-1758.

[14] Wong, C.Y., et al.,"A Real-time Sub-carrier Allocation Scheme for Multiple Access Downlink OFDM Transmission," *Proc. IEEE VTC,* Fall, pp. 1124-1128, 1999.

[15] Rhee, W., and Cioffi, J.M., "Increase in Capacity of Multiuser OFDM System Using Dynamic Subchannel Allocation," *Proc. IEEE VTC,* Spring, pp. 1085-1089, 2000, Tokyo, Japan.

[16] Yin, H., and Liu, H., "An Efficient Multiuser Loading Algorithm for OFDM-based Broadband Wireless Systems," *Proc. IEEE Globecom,* pp. 103-107, 2000.

[17] Kivanc, D., and Liu, H., "Subcarrier Allocation and Power Control for OFDMA," *Proc. 34th Asilomar Conference on Signals, Systems and Computers,* pp. 147-151, November 2000.

[18] Kivanc, D., Li, G., and Liu, H., "Computationally Efficient Bandwidth Allocation and Power Control for OFDMA," *IEEE Transactions on Wireless Communications,* November 2003, volume 2, pp. 1150-1158.

[19] Kim, I., et al., On the Use of Linear Programming for Dynamic Subchannel and Bit Allocation in Multiuser OFDM," *Proc. IEEE Globecom,* pp. 3648-3652, 2001.

[20] Pfletschinger, S., Munz, G., and Speidel, J., "Efficient Subcarrier Allocation for Multiple Access in OFDM Systems," *Proc. Int'l OFDM Workshop,* 2002, Hamburg, Germany.

[21] Pietrzyk, S., and Janssen, G.J.M., "Multiuser Subcarrier Allocation for QoS Provision in the OFDMA Systems," *Proc. IEEE VTC,* Fall, pp. 1077-1081, September 2002, Vancouver, Canada.

[22] Krunz, M., "Bandwidth Allocation Strategies for Transporting Variable-Bit-Rate Video Traffic," *IEEE Communications Magazine,* January 1999, pp. 40-46.

[23] Le Bodic, G., Irvine, J., and Dunlop, J., "Resource Cost and QoS Achievement in a Contracts-based Resource Manager for Mobile Communications Systems," *Proc. Eurocomm,* Munich, 2000.

[24] Cioffi, J.M., "A Multicarrier Primer," Amati Communications Corporation and Stanford University, November 1991.

[25] Proakis, J.G., *Digital Communications,* McGraw-Hill, 1995.

[26] Zhang, Y., and Letaief, K.B., "Multiuser Subcarrier and Bit Allocation along with Adaptive Cell Selection for OFDM Transmission," *Proc. IEEE ICC,* pp. 861-865, 2002.

[27] Li, J., et al., "A Novel Broadband Wireless OFDMA Scheme for Downlink in Cellular Communications," *Proc. IEEE WCNC,* pp. 1907-1911, 2003.

[28] Kuhn, H.W., "The Hungarian Method for the Assignment Problem," *Naval Research Logistics Quarterly* 2, 1955.

[29] Dondeti, V.R., and Emmons, H., "Max-Min Matching Problems with Multiple Assignments," *Journal of Optimization Theory and Applications,* November 1996, volume 91, pp. 491-511.

[30] Keller, T., and Hanzo, L., "Adaptive Multicarrier Modulation: A Convenient Framework for Time-Frequency Processing in Wireless Communications," *IEEE Proceedings of the IEEE,* May 2000, volume 88, pp. 611-640.

[31] Lai, S.K., et al., "Adaptive Tracking of Optimal Bit and Power Allocation for OFDM Systems in Time-Varying Channels," *Proc. IEEE WCNC'99,* pp. 776-780, 1999.

Chapter 6

RRM algorithms for OFDMA with intra-cell spectrum reuse

In the downlink OFDMA system, a subcarrier may not only be of high quality to one user while being of low quality to another user but it may well be of high quality to a number of users. In such a case, it may be power efficient to reuse this subcarrier within a cell in order to address multiple users by stacking multiple signals on this subcarrier. In this way, an embedded constellation is created where constellation symbols of one link are centered around the constellation symbols of another link, thus increasing the number of available independent links and extending it over the limit determined by the number of available subcarriers.

This chapter considers a downlink OFDMA system with intra-cell reuse of subcarriers by applying embedded modulation. Relations are described that allow to design the embedded signal constellation. This is followed by the presentation of RRM algorithms for subcarrier assignment, bit loading, and power allocation for the downlink OFDMA system reusing subcarriers within a cell. This chapter is an extension of our previous work based on [1].

This chapter is organized as follows. Section 6.1 provides a literature review of research work on RRM schemes for single-cell OFDMA systems with intra-cell reuse of subcarriers. Since this research area is relatively new and unexplored, we also present other work that constitutes the context of this area, as for example chosen papers in the broadcasting domain. In Section 6.2 we present a simplified system model, where reuse of subcarriers is applied for the uplink. Section 6.3 introduces the considered system model and formulates the allocation problem. Section 6.4 presents a set of state-of-the-art algorithms efficiently distributing resources in the downlink OFDMA system with intra-cell spectrum reuse. Section

6.5 verifies the performance of these algorithms by simulations. Finally, a summary and conclusions are given in Section 6.6.

6.1 LITERATURE REVIEW

The idea of superimposing two information streams (high-rate on low-rate) was suggested and studied by Cover in the information theory setting of broadcast channels in [2]. It was shown that higher joint rates can be achieved by superimposing information streams than by using time multiplexing with the same transmit power.

This idea was later exploited in the high definition television (HDTV) scenario [3, 4] by Ramchandran et al. to allow for multiresolution transmission by applying so-called *embedded modulation*. In such a system the detailed information signal (higher resolution) referred to as "satellite" is embedded in the "base" information signal (lower resolution) referred to as "cloud." Then the receivers with a good channel (i.e., located close to the HDTV transmitter) can decode both signals and thus can benefit from a higher reception quality, while those with a bad channel can decode only the signal corresponding to the lower quality. Also, the application of successive detection and subtraction was suggested to distinguish the two component signals, where the receiver first decodes the "cloud," subtracts the result from the received signal and subsequently decodes the "satellite" [5]. The proposal posed no restrictions either on the form or on the size of the component constellations. However, it considered only two embedded (stacked) signals. Later, in [6, 7], Pradhan and Ramchandran applied the embedded modulation idea into multicarrier systems in order to realize layered data transmission. The optimization problem for the two component signals was to maximize the sum-rate subject to the total power constraint. The various layers correspond to various levels of importance of data streams (i.e., having different BER requirements), and they are not related to two users having different channel conditions. Thus no multiuser diversity is considered.

In [8], Kim and Stuber verified the performance of the multiresolution OFDM system for HDTV application. They focused on the influence of intercarrier interference (ICI) for a system where only two resolution levels are considered.

In [9], Cheng and Verdu considered the extension of a single user water-filling to a multiuser (two-user) water-filling in Gaussian multiaccess channels with ISI. Also here, one signal is allowed to be superimposed onto another signal. Multiuser diversity is considered here, since the two signals are intended for two users having different channels. The objective of two-user water-filling is to maximize the sum-rate by applying optimal power and bit loading. The superimposed signals may be separated by successive decoding and subtraction. Also, the idea of "successive water-filling" is suggested, where each *successive* user does water-filling on top of the previous user's signal plus noise. The user with the highest priority (i.e., the first user) does water-filling only on top of the noise. The other users must take into account the signals of the previous users. In [10], Diggavi addressed a similar problem of multiuser water-filling explicitly for a multicarrier system. The optimization goal is to maximize the sum-rate subject to individual power constraints. Only two users with equal priorities are considered. The complexity of the optimal algorithm increases exponentially with the number of users. The reception in this proposal is also based on successive detection and is realized using a so-called "onion-peeling" approach. In [11], Zeng et al. presented a water-filling algorithm, which is more computationally efficient than that in [10] and which solves the problem of maximizing the sum rate of two users. From the simulations it can be seen that for equal-priority users the optimal solution is given by applying FDMA, while for unequal-priority users the FDMA scheme does not achieve the optimal solution and the two users should share some subcarriers. In [12] Ohno et al. confirmed that "sum-rate optimal multicarrier users could share one or more subcarriers." Also here, the optimization problem is to maximize the sum rate.

In [13], Hoo et al. formulated the optimization problem with the goal to minimize the total power consumption subject to individual data rates and bit error probabilities required by the users, considering in this way, QoS support. In addition, the authors addressed the issue of signal constellation design by applying embedded modulation.

The idea of signal superposition was considered by Sari et al. in [14–17] with the intention to increase the number of simultaneously served users. Here, in the general scenario of various multiple access schemes (CDMA, OFDMA, etc.) the idea is to first distribute resources in a nonsharing (orthogonal) manner so only one user is allowed to exploit a carrier waveform (e.g., frequency channel, code) and then, if all the waveforms are distributed, to allow for allocation with waveform sharing. Two layers of superposition are considered, which results in a potential doubling of the original system capacity. The users are distinguished by the use of multiuser detection techniques (e.g., joint detection or successive detection).

The idea of distinguishing signals in another domain (i.e., other than time or frequency) was also studied in [18] by Mazzini. Here, the uplink signals originating from different users are distinguished thanks to appropriate power separations hence the name of the proposed scheme: power division multiple access.

In [19], Janssen considered the problem of constellation design for superimposed signals used to address multiple users in the downlink scenario. The component signals are BPSK or QPSK and the resulting embedded constellation is referred to as compound QAM. In [19], the required relations between the phases and amplitudes of the component signals are established, taking into account users' channel gains. Also, the uplink scenario is considered where random phases of signals originating from different users are assumed. There is no limitation on the number of superimposed signals like in the papers on embedded modulation [3, 4, 6, 7]. It is shown that it is relatively cheap (in terms of transmit power) to superimpose one user's signal on another provided that sufficiently different channel qualities are available.

In [1], Pietrzyk and Janssen considered a problem of radio resource allocation for OFDMA systems reusing subcarriers within a single cell (intra-cell reuse) by the application of embedded modulation, which allows to stack signals addressed to various users. A low-complexity, time-scalable algorithm is proposed, which allows for efficient allocation of subcarriers and transmit power according to individual user traffic requirements. It is shown that signal stacking is more power efficient than addressing users with a higher order modulation and applying for example time sharing.

Summarizing the work done in the area of OFDMA with intra-cell reuse of subcarriers, although many papers exist, very few actually present the algorithms for radio resource allocation for multiple access systems. Many papers approach the problem from the information-theory perspective, where the aim is to maximize the sum-rate of a system. Although such an approach allows for finding a Shannon-like maximal system capacity, it is not very practical from a communication system perspective because sum-rate maximization may cause some users to get higher data rates than they really require, while others may be left without any assignment. Also, most of the considered research work does not address the issue of signal constellation design and/or they expose the solutions only for the two-signal case. In the next sections we make an attempt to fill these gaps.

6.2 REUSE OF SUBCARRIERS FOR A SIMPLIFIED UPLINK MODEL

Before considering in detail the target downlink OFDMA system with intra-cell reuse, we shall address a simplified system model, where reuse of subcarriers is applied for the uplink for the purpose of congestion control. This model will allow to better understand more complex schemes for the downlink.

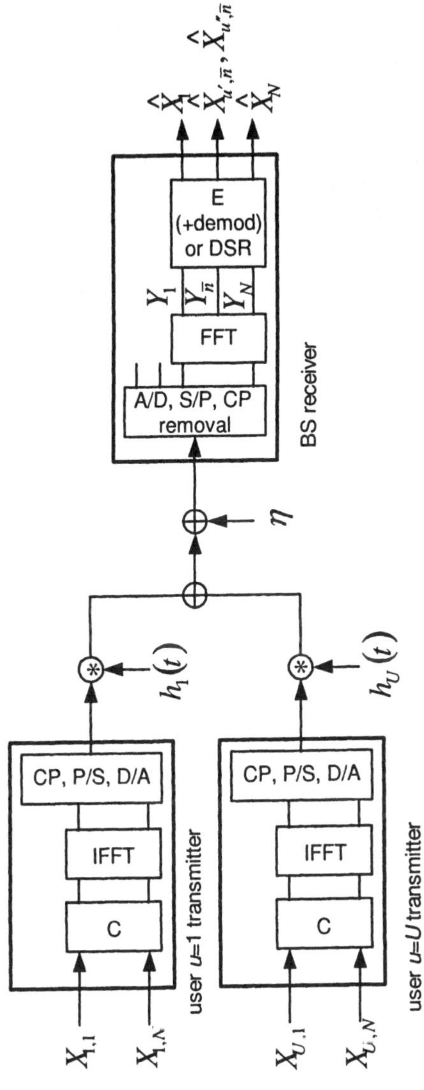

Figure 6.1 System model for the uplink of a collision-based OFDMA system employing *embedded* modulation.

A congestion state, defined as a situation when all the radio resources (here OFDM subcarriers) are already distributed but still there are users to be served, may happen occasionally even to a well-designed wireless access system. In such a situation, conventional systems (GSM, GPRS, [20]) simply block or at most queue the *excessive* demands. The negative effect of such an approach is especially visible during a handover to a congested cell, where, if no countermeasures are applied, the ongoing call is dropped. Following the "soft QoS" approach [21], we allow the excessive users to the system [14, 16]. In the OFDMA context, this is achieved by letting multiple users (here we restrict to two users only for simplicity of exposition) to simultaneously utilize (i.e., to share) the same set of OFDM subcarriers. In order to resolve the resulting collisions, we choose such component signals of the two users so that the *embedded* constellation is formed and therefore both users can be well detected at the base station receiver. The ideas introduced in this section reveal a potential for stacking several signals aimed for different users on the same OFDMA subcarrier in order to increase system capacity and / or increase power efficiency. They are followed further in this chapter for the downlink scenario.

The remainder of this section is organized as follows. In Subsection 6.2.1 we briefly present the considered system model. In Subsection 6.2.2 we describe three allocation methods used to appoint users and subcarriers for collision. In Subsection 6.2.3 the results of numerical experiments are presented. Conclusions are given in Subsection 6.2.4.

6.2.1 System model

The considered multiple access system consists of U user terminals communicating simultaneously with one base station (BS) over a frequency selective fading channel, as shown in Figure 6.1. Symbols belonging to user u are assigned to subcarriers according to the allocation matrix $\boldsymbol{C} = \{C_{u,n}\}$. Subsequently, the symbols are OFDM-modulated, a cyclic prefix (CP) is added, and the symbols are converted into a serial stream. The BS receives a mixture of individual users' signals convolved with the appropriate channel $h_u(t)$ plus an AWGN term. After inverting the transmitter operations, assuming a perfect guard from the channel echoes by the CP and perfect synchronization, the nth subcarrier signal received at the BS can be expressed by

$$y_n = \sum_{u=1}^{U} C_{u,n} X_{u,n} H_{u,n} + \eta, \tag{6.1}$$

where: $u = 1, \ldots, U$ and $n = 1, \ldots, N$ indicate user u and subcarrier n, respectively; the element $C_{u,n}$ of the subcarrier allocation matrix $\boldsymbol{C}$ indicates

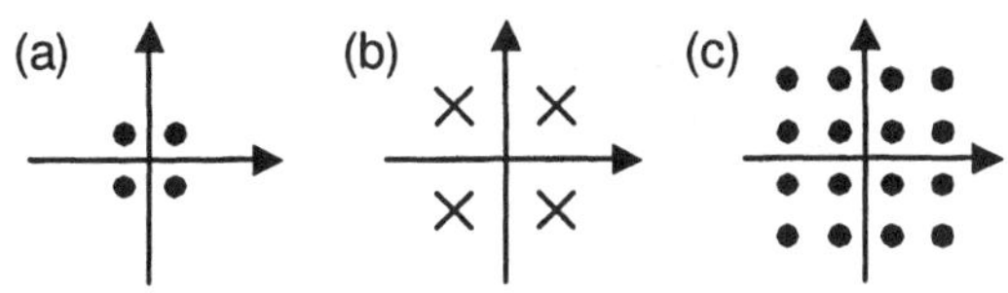

Figure 6.2 Example of 4-QAM constellations: (a) component minor, (b) component major, and (c) resulting 16-QAM embedded.

whether the subcarrier n is allocated to user u ($C_{u,n} = 1$ meaning that n is allocated to u, $C_{u,n} = 0$ meaning that n is not allocated to u); the data symbol $X_{u,n}$ of the uth user transmitted using nth subcarrier; $H_{u,n}$ is the channel frequency response of the uth user channel $h_{u,n}(t)$ at the nth subcarrier; and η is AWGN.

6.2.2 Collision allocation methods

Let us consider a pair of users u' and u'' allocated the same subcarrier $\overline{n}$ and thus being in collision (interfering each other). In order to resolve a collision, Sari in [14] proposed that one user shall transmit a 4-QAM signal (called the *major* signal) with the amplitude two times larger than the amplitude of a 4-QAM signal (called the *minor* signal) of another user. Additionally, both signals should be in-phase when present at the BS receiver. Then, assuming the same channel conditions for both users, the resulting signal, which is a sum of the component signals, has the symbols from the alphabet of 16-QAM constellation, as can be seen in Figure 6.2, and hence both component signals can be distinguished. The resulting 16-QAM constellation can be referred to as an *embedded* constellation, which is built out of the two competent 4-QAM signals. In the considered system each user experiences different channel gains, which in addition are highly variable in the frequency domain. This is different than the case described in [14], where channel conditions are assumed the same for both users. Therefore, we generalize Sari's idea [14], and consider a collision-based system, which not only takes into account the channel variability but also exploits this phenomenon to match the component signals. This is realized by designating for a collision such subcarrier for which the component signals (two 4-QAM), after passing the channel, are seen by the BS in a form of a desired, embedded constellation.

Another possibility to distinguish the two component 4-QAM signals, besides detecting the 16-QAM, is to apply the dual signal receiver [22], which is a narrow-band multiuser receiver based on successive detection and cancellation. At the BS, after the FFT-based OFDM demodulation, the signals Y_n are fed into a bank

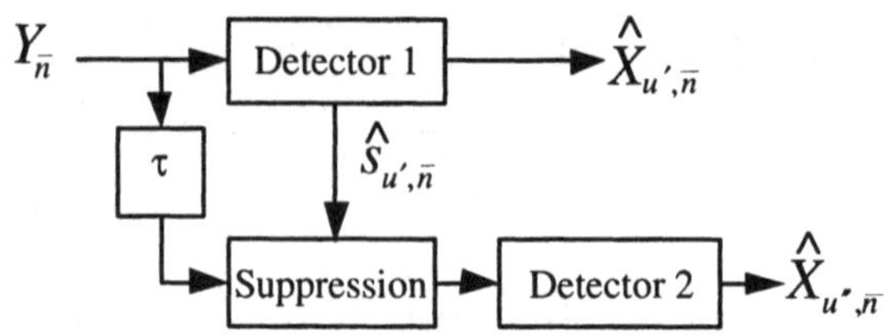

Figure 6.3 Dual-signal receiver applied to resolve a collision on $\overline{n}$th subcarrier.

of parallel per-subcarrier DSRs (see Figure 6.1). Each can distinguish up to two cochannel signals. This is explained in the following.

After CP removal and OFDM demodulation, the $\overline{n}$th subcarrier signal received at the BS can be expressed by

$$Y_{\overline{n}} = X_{u',\overline{n}} H_{u',\overline{n}} + X_{u'',\overline{n}} H_{u'',\overline{n}} + \eta. \tag{6.2}$$

Assume that $X_{u',\overline{n}} H_{u',\overline{n}}$ is the major and $X_{u'',\overline{n}} H_{u'',\overline{n}}$ is the minor signal. The DSR, consisting of two detectors in succession as depicted in Figure 6.3, is capable of detecting both colliding signals in the following manner. Detector 1 locks onto the major signal and detects $\hat{X}_{u',\overline{n}}$. In addition, Detector 1 estimates the signal $\hat{s}_{u',\overline{n}}$, which is used to suppress the major signal in a τ delayed version of the input signal $Y_{\overline{n}}$. After suppression of the major signal, Detector 2 locks onto the minor signal and detects $\hat{X}_{u'',\overline{n}}$. In such an iterative way, both component signals are detected. It should be noted however, that though the component signals do not have to be in-phase, there should be a sufficient power separation between them. A detailed study of the DSR can be found in [22].

In the following, we consider three realizations of the collision-based uplink OFDMA system:

- Detection of 16-QAM constellation with the application of preequalizer;
- Detection of 16-QAM constellation with the application of post-equalizer;
- Detection with the application of dual-signal receiver.

6.2.2.1 System with preequalizer

Here the OFDMA system with preequalization is considered. In Figure 6.1, the preequalizer is realized by the use of subcarrier allocation matrix $\boldsymbol{C}$, which, instead of containing only zeros and ones, contains complex preequalization coefficients;

the post-equalizer matrix $\boldsymbol{E}$ and bank of DSRs at the base station become unnecessary in this case.

With the use of the preequalizer, each component signal is independently predistorted prior to IFFT. Thus, the two colliding signals can be matched to each other in order to form a "desired" constellation and to prepare for the transmission over a multipath channel. After CP removal and OFDM demodulation, the $\overline{n}$th subcarrier signal received at the BS can be expressed by

$$Y_{\overline{n}} = C_{u',\overline{n}}X_{u',\overline{n}}H_{u',\overline{n}} + C_{u'',\overline{n}}X_{u'',\overline{n}}H_{u'',\overline{n}} + \eta. \tag{6.3}$$

where $C_{u',\overline{n}}$ and $C_{u'',\overline{n}}$ are here the preequalization coefficients for the two colliding signals.

In order to determine $C_{u',\overline{n}}$ and $C_{u'',\overline{n}}$ we need to know the channels and the desired constellation. We define the desired constellation as the one that can be easily composed out of the two other constellations and then detected at the receiver. In order to realize the desired resulting constellation, the component signals should be in a certain relation ζ (which is a complex number) to each other. For 4-QAM signals considered by Sari, one component signal takes its real and imaginary values from the alphabet $\{-2, +2\}$ and another from $\{-1, +1\}$, which results in a relation factor $\zeta = 2$. This means that for the same constellation point, the amplitude of one signal is twice the amplitude of another signal and both signals are in-phase. Given $H_{u',\overline{n}}$, $H_{u'',\overline{n}}$ and ζ, the preequalization values can be expressed by the following relations

$$C_{u',\overline{n}} = H^{-1}_{u',\overline{n}}, \tag{6.4}$$

and

$$C_{u'',\overline{n}} = \zeta H^{-1}_{u'',\overline{n}}. \tag{6.5}$$

The important aspect now is how to choose the subcarrier $\overline{n}$, where users u' and u'' should collide. Simple methods include sequential [14] or random allocations. In sequential allocation, at the time instant $t = 1$, the collision is placed on a subcarrier $n = 1$ shared among users $u = 1$ and $u = 2$. Subsequently, at the time instant $t = 2$, the subcarrier $n = 2$ is shared among users $u = 2$ and $u = 3$, and so on. In random allocation, at the time instant $t = 1$ the subcarrier $n = 1$ is shared among two randomly chosen users. For both methods, the process continues until each subcarrier is used for the collision.

Both sequential and random allocation suffer from high power penalty required to preequalize these subcarriers, which are in deep fades, because the collision allocation is performed regardless of the channel conditions. Another possibility is to allocate a collision in such a way that the sum transmit power of the involved users is minimized. The allocation problem can be formulated as follows: given $H_{u,n}$ and ζ find such u', u'' and $\overline{n}$ so that

$$\Phi = \langle C_{u',\overline{n}} X_{u',\overline{n}} \rangle + \langle C_{u'',\overline{n}} X_{u'',\overline{n}} \rangle \tag{6.6}$$

is minimized. In the above formula $\langle x \rangle$ is the power of the waveform x. The above problem boils down to allocating a collision on such u', u'', and $\overline{n}$, which experience best channel conditions. The main drawback here is the need to know apriori the channel conditions for both users.

6.2.2.2 System with post-equalizer

Here, the OFDMA system with post-equalization is considered. In Figure 6.1, the post-equalizer is realized by using matrix $\boldsymbol{E}$, which is present at the base station. In this case, the subcarrier allocation matrix $\boldsymbol{C}$, indicates only the subcarrier allocation (it does not act as a preequalizer) and the bank of DSRs is unnecessary.

With the use of post-equalization, the component signals are post-distorted (i.e., corrected) jointly at the base station. After CP removal and OFDM demodulation, the $\overline{n}$th subcarrier signal at the BS can be expressed by

$$Y_{\overline{n}} = E_{\overline{n}}(X_{u',\overline{n}} H_{u',\overline{n}} + X_{u'',\overline{n}} H_{u'',\overline{n}} + \eta), \tag{6.7}$$

where $E_{\overline{n}}$ is a post-equalization coefficient, which belongs to the post-equalizer matrix $\boldsymbol{E}$

$$\boldsymbol{E} = \begin{bmatrix} E_1 \ E_2 \ \dots \ E_N \end{bmatrix}^T. \tag{6.8}$$

Because the two colliding signals cannot be matched to each other at the transmitter as it was in the previous case, allocation should be done on such u', u'', and $\overline{n}$ for which the component signals after passing the channel are in a desired proportion. The allocation problem can be formulated as follows: given $H_{u,n}$ and ζ find u', u'' and $\overline{n}$ so that

$$\Delta = abs(H_{u',\overline{n}} - \zeta H_{u'',\overline{n}}) \tag{6.9}$$

is minimized. In the above formula $abs(x)$ is the absolute value of x. In such way frequency selectivity of a multipath channel is exploited to match the component signals.

The post-equalization coefficient is determined based on the channel corresponding to the minor signal as expressed by

$$E_{\overline{n}} = H^{-1}_{u'',\overline{n}}. \tag{6.10}$$

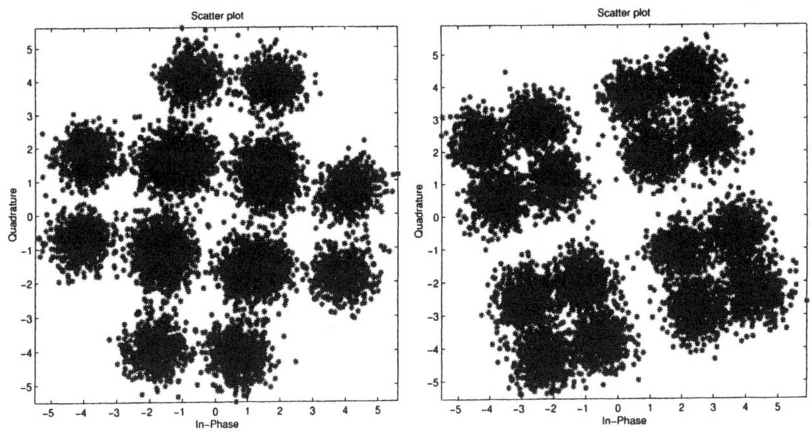

Figure 6.4 Example of insufficient $SIR = 4$ dB (left-side) and sufficient $SIR = 8$ dB (right-side) separating the two component 4-QAM signals.

A disadvantage of post-equalization is that in the case of low-quality subcarriers, the post-equalizer amplifies not only the wanted signal but also the noise. Therefore, means of allocating collisions only for users and subcarriers with good quality channels can be considered. In the simulations presented further in this section, 10% of the worst subcarriers are discarded for each user, and then collision is allocated to the chosen remaining subcarrier. Also in this case, the channel conditions of both users should be known apriori in order to find good collision allocation.

6.2.2.3 System with dual-signal receiver

Here, we consider the OFDMA system applying the dual-signal receiver (DSR) to distinguish the component signals. In Figure 6.1, at the BS a bank of per-subcarrier DSRs is present ($\boldsymbol{E}$ matrix is unnecessary). Different allocations of users for collision on a single subcarrier may result in different embedded constellations present at the input of DSR. Figure 6.4 depicts two example cases of such an allocation. The difference between the left-hand and the right-hand picture stems from the fact that the BER performance of the minor signal is strongly determined by the signal-to-interference ratio (SIR), which is the ratio of the power of the major to the power of the minor signal. For the left-hand picture of Figure 6.4 SIR is not sufficiently shielding the minor signal, while in the right-hand picture the SIR separation is large enough. As can also be seen in Figure 6.5, for $SIR \geq$

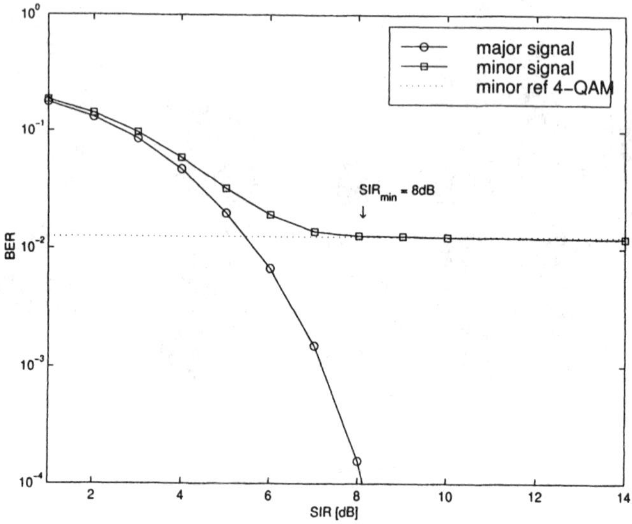

Figure 6.5 Performance of the minor and the major signal as a function of SIR. Assumed minor signal $E_b/N_0 = 4$ dB.

$SIR_{min} \simeq 8$ dB, the difference in the BER performance between the minor signal and a corresponding ideal 4-QAM signal is negligible. Therefore, the collision should be allocated in such a way that: (1) the SIR fulfills the required condition in order to assure good separation between two colliding signals, and (2) the quality of the minor signal subcarrier is maximized. The allocation problem can be formulated as follows: given $H_{u,n}$ and SIR_{min} find such u', u'' and $\overline{n}$ so that

$$SIR = \frac{\langle X_{u',\overline{n}} H_{u',\overline{n}} \rangle}{\langle X_{u'',\overline{n}} H_{u'',\overline{n}} \rangle} \geq SIR_{min}, \tag{6.11}$$

and

$$SNR = \frac{\langle X_{u'',\overline{n}} H_{u'',\overline{n}} \rangle}{\langle \eta \rangle} \tag{6.12}$$

is maximized (or is larger than a certain threshold). The above problem boils down to searching pairs of subcarriers fulfilling the SIR condition and, subsequently, finding among them the one that experiences the best minor signal-to-noise ratio.

Similar algorithms using such power conditions have been proposed in [23]. However, in [23] the variations of SIR were obtained by exploiting various path

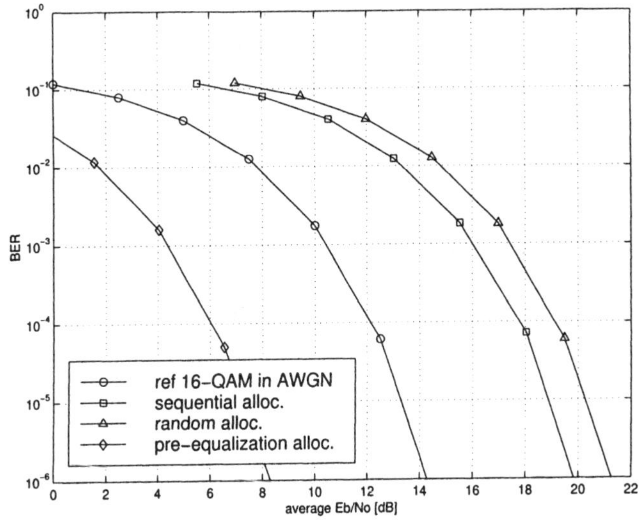

Figure 6.6 Performance comparison of various collision-allocation algorithms for a system with preequalizer. BER is given for the resulting 16-QAM embedded constellation (thus performance of 16-QAM is given as a reference).

loss and shadowing profiles experienced by different users. Here, for this purpose, frequency selective fading is employed solely.

6.2.3 Simulation results

In order to evaluate the performance of the introduced collision allocation algorithms, we performed Monte Carlo simulations using the system parameters as introduced in Chapter 4.

6.2.3.1 Performance of collision allocation algorithms for a system with preequalizer

The performance of the collision allocation method for a system with preequalizer in comparison to sequential and random allocation, as well as to reference 16-QAM, is shown in Figure 6.6.

It can be concluded that in the case of a system with preequalizer, good BER performance stems from choosing best-quality subcarriers for collision and perfect

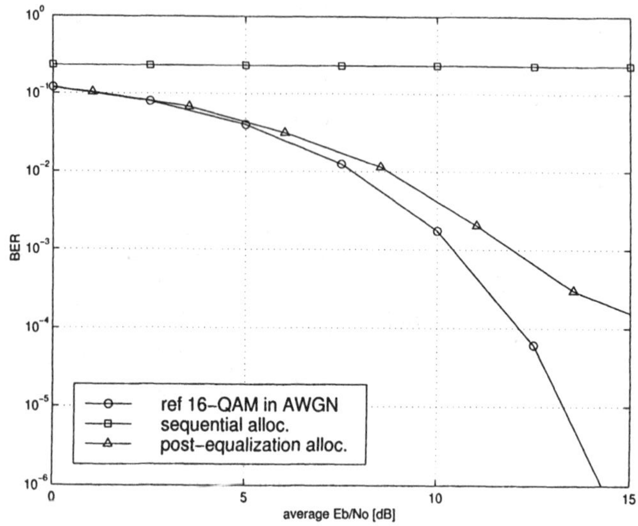

Figure 6.7 Performance comparison of various collision-allocation algorithms for a system with post-equalizer. BER is given for the resulting 16-QAM embedded constellation (thus performance of 16-QAM is given as a reference).

matching of the component signals due to the application of preequalization. This method, however, could be problematic for implementation in the uplink, where channel probing and controlling all users' preequalization coefficients would be a complex and signaling-heavy task. On the other hand, application of this approach for the downlink seems to be less problematic since it is easier to coordinate signals aimed for multiple users when they originate from the same place (i.e., the base station). The idea of preequalization of signals for the downlink transmission lays the basis for RRM algorithms introduced further in this chapter.

6.2.3.2 Performance of collision allocation algorithms for a system with post-equalizer

The performance of the collision allocation method for a system with post-equalizer in comparison to sequential allocation and reference 16-QAM is shown in Figure 6.7. We can observe that in the case of a system with post-equalizer, it is of primary importance to choose channels that "predistorts" the users' signals so that they

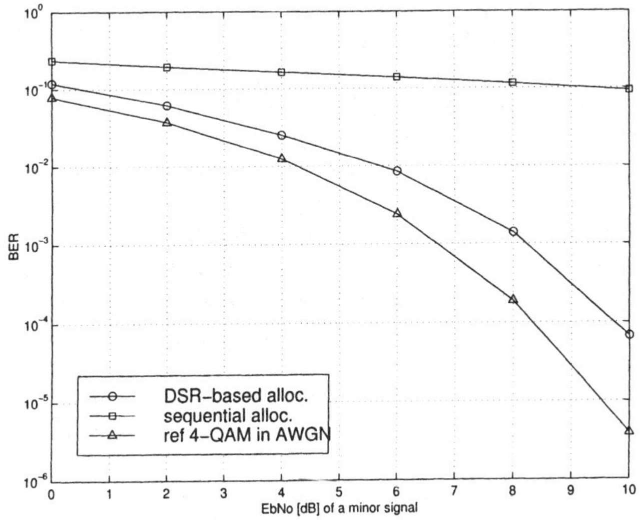

Figure 6.8 Performance of the allocation algorithm for dual-signal receiver with sequential allocation. BER is estimated for the minor 4-QAM signal only (thus performance of 4-QAM is given as a reference).

arrive at the receiver in a form of the desired, embedded constellation. Nevertheless, choosing good-quality channels also helps here.

This approach may be considered for the application to the uplink of a collision-based OFDMA system with fixed transmit power per subcarrier. In the case of such systems the user transmitter is of low complexity, while the base station takes care of detection of the resulting embedded constellation and of proper choice of users and subcarrier for collision. This idea, however, is not further followed in this book, since all the considered system cases allow for transmit power adaptation.

6.2.3.3 Performance of collision allocation algorithms for a system with dual-signal receiver

The major signal is always detected in the presence of an interfering minor signal. Assuming a perfect major signal detection and a proper separation of the two signals, the minor signal is virtually interference-free. Therefore, in a congestion situation, the minor signal is a better candidate for already established (existing) calls or for calls with a guaranteed QoS. This is because the BER performance of the minor signal is the same before and after the collision. Thus, in the following we focus on the performance of the minor signal.

The performance of the described collision allocation method for a system with dual-signal receiver is depicted in Figure 6.8. As already mentioned, we consider BER of only the 4-QAM signal corresponding to the minor signal. It can be observed that also in this case the presented method outperforms sequential allocation and is only slightly worse than the reference 4-QAM. DSR can be applied both for the uplink and for the downlink to distinguish the component signals. Its application for the uplink may be motivated by the fact that DSR does not require the users' signals to be preprocessed at the transmitters. Instead, independent signals may be combined assuming there is a sufficient power separation between them. However, instructing mobile terminals to use particular subcarriers for collision requires a control channel from the base station to the users. In the downlink, we would preferably like to control the power and the phase of the transmitted signals so the collision-allocation algorithm considered here is not suitable. Nevertheless, a dual-signal receiver may be used at the mobile terminal to realize the detection of the embedded constellation by applying the onion-peeling principle. This idea, with the generalization of DSR to allow for detection of more than two signals, is further followed in this chapter.

6.2.4 Conclusions for a simplified uplink model

In this section we have considered methods for congestion resolving in the uplink OFDMA scenario. This approach allows extra users to access the system in a collision-based manner when all subcarriers are already allocated.

We analyzed three cases of collision-based channel allocation for the uplink of an OFDMA system that includes systems applying: preequalization, postequalization, or dual-signal receiver. In all of the considered cases, the presented algorithms show better performance than the sequential methods performing allocations regardless of underlying channel conditions.

The system with preequalizer is a straightforward extension of Sari's idea [14]. Its superior performance stems from the ability to allocate collision on best quality subcarrier. The idea of predistorting signals at the transmitter, as applied for the downlink case, is further elaborated in this chapter. For the system with postequalizer, shifting part of the processing requirements (equalization) to the base station simplifies the user terminal, which may be considered an advantage. Here, the channel selection is truly used to match the two colliding signals. However, this system exhibits the worst performance and is not considered further in this book.

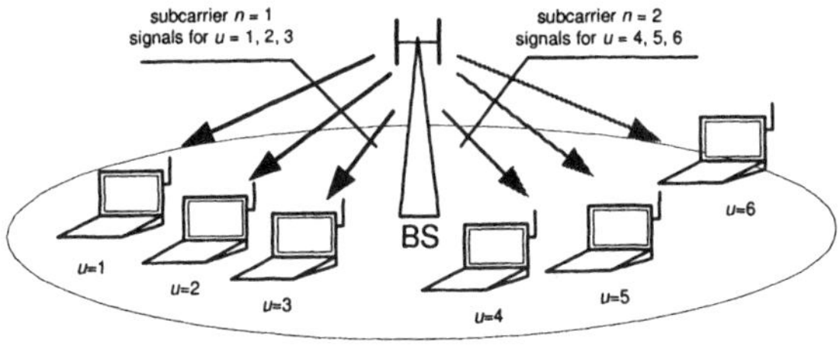

Figure 6.9 Single-cell downlink OFDMA system with embedded modulation and intra-cell reuse of subcarriers. Example for $U = 6$ users, $N = 2$ subcarriers and $L = 3$ M-QAM signals on each subcarrier.

The system applying dual-signal receiver has an advantage of not requiring either phase or power corrections. The price is a more complex receiver at the base station, which must be equipped with a bank of per-subcarrier DSRs.

Summarizing, the key ideas that are further followed in this chapter include: (a) applying collision-based allocation (i.e., stacking signals addressed for multiple users on the same subcarrier) in order to increase the system capacity, and (b) using best-quality subcarriers for collision allocation. The major drawback of the presented methods is the consideration of only two users and one subcarrier for collision, where embedded constellation can be formed. Such approach is justified only for the temporal congestion situation to allow for serving the excessive traffic, where reallocation of the resources for the existing traffic may be too costly.

6.3 DOWNLINK SYSTEM MODEL AND PROBLEM FORMULATION

In this section, a model of a single-cell downlink OFDMA system with intra-cell reuse of subcarriers is presented. This is followed by a formulation of the radio resource allocation problem applicable to this system model.

6.3.1 System model

The single-cell downlink OFDMA model as presented in Chapter 5 is applied here. The system consists of one BS serving U users using N OFDM subcarriers over a total bandwidth of BW, as shown in Figure 6.9. The CSI is characterized by the channel gains $\boldsymbol{G} = [G_{u,n}]$ and AWGN variance σ^2. Also, the QoS model introduced in Chapter 5 is applied.

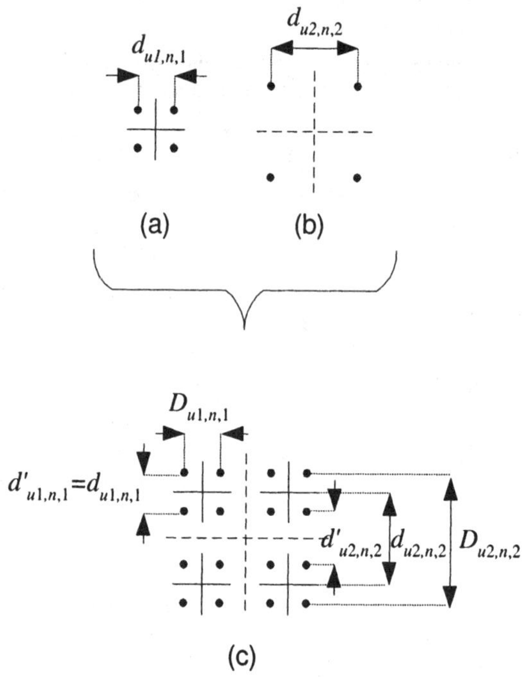

Figure 6.10 Example embedded 16-QAM constellation (c) composed of stacking $L = 2$ 4-QAM signals (a) and (b) on a subcarrier n. The constellation design parameters for subcarrier n are: $d'_{u,n,l}$ - distance between constellation points, $d_{u,n,l}$ - distance between constellation symbols, $D_{u,n,l}$ - constellation width. Each of these 4-QAM signals may be addressed for different users $u1, u2$.

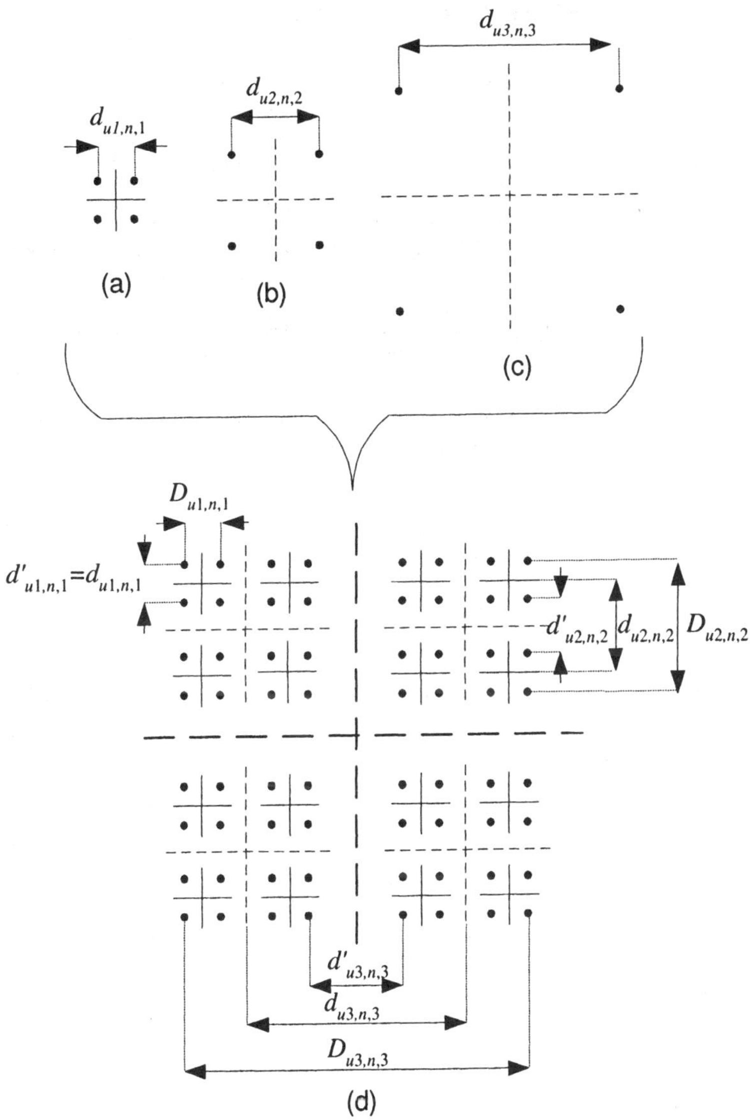

Figure 6.11 Example embedded 64-QAM constellation (d) composed of stacking $L = 3$ 4-QAM signals (a), (b), and (c) on a subcarrier n. Each of these 4-QAM signals may be addressed for different users $u1, u2, u3$.

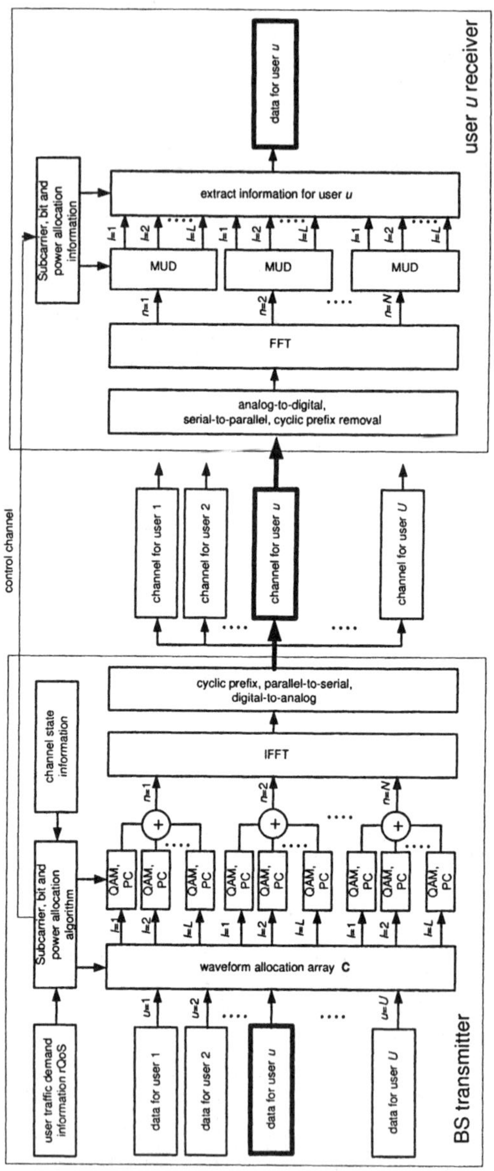

Figure 6.12 Block model of a single-cell downlink OFDMA system with embedded modulation and intra-cell reuse of subcarriers. (*After:* S. Pietrzyk and G.J.M. Janssen, 2004 [1].)

The main difference between the system model considered in this chapter and the one presented in the previous chapter is the fact that here a subcarrier can be assigned to multiple users within a cell (thus spectrum is reused), while in Chapter 5 it could be allocated exclusively to one user only. As a result, in the case of the OFDMA system with reuse of subcarriers within a cell, embedded constellation is present on each subcarrier [4, 10, 13, 19]. This embedded constellation is created by hierarchical stacking of up to L, M-QAM symbols (also referred to as signals) each indicated by $l = 1, \ldots, L$. Each M-QAM symbol carries $b = \log_2(M)$ bits. In order to keep the resulting number of states in the embedded constellation within practical limits, we assume that $L = 3$ and $M = 4$. Of course the presented analysis holds also for $L > 3$ and $M > 4$. Effectively, the resulting embedded constellation can take the form of QPSK (one 4-QAM symbol), 16-QAM (two stacked 4-QAM symbols) as shown in Figure 6.10, or 64-QAM (three stacked 4-QAM symbols) as shown in Figure 6.11.

Let us define a (carrier) waveform as a pair (n, l) of a subcarrier n and a single 4-QAM signal l located on this subcarrier. Using only one waveform we can create a single separable link over the air interface. Allocation of waveforms to users is controlled at the BS with the use of the allocation array $\boldsymbol{C} = [C_{u,n,l}]$, where $C_{u,n,l} = 1$ indicates that waveform (n, l) is allocated to user u, while $C_{u,n,l} = 0$ indicates that waveform (n, l) is not allocated to user u. It should be emphasized that waveforms on the same subcarrier can be allocated to different users, thus realizing the intra-cell subcarriers' reuse idea, which can be seen in Figure 6.11 (users $u1, u2, u3$).

The block model of the considered downlink OFDMA system with embedded modulation is depicted in Figure 6.12. At the BS, after allocation to users, each waveform (n, l) is 4-QAM modulated and power controlled. Let us define the average power of a waveform (n, l) when allocated to user u as follows [24]

$$P_{u,n,l} = \frac{d_{u,n,l}^2(M-1)}{6}, \tag{6.13}$$

where M is left on purpose as a variable to make this relation universal, $M = 2^b$ is some power of 4 (b is an even integer), $d_{u,n,l}$ is the distance between the constellation symbols. For the first waveform $(n, 1)$ on a subcarrier n, the distance between the constellation symbols $d_{u,n,1}$ is equal to the distance between the constellation points and can be expressed by

$$d_{u,n,1} = d'_{u,n,1}. \tag{6.14}$$

The distance $d'_{u,n,l}$ between the constellation points can be expressed by

$$d'_{u,n,l} = \sqrt{6T_{u,n}^{-1}}, \tag{6.15}$$

with

$$T_{u,n} = \frac{G_{u,n}^2}{2\sigma^2\Gamma_u}, \tag{6.16}$$

where $T_{u,n}$ is the normalized channel gain, while Γ_u is the SNR gap that can be expressed by the following relation [24]

$$\Gamma_u = \frac{1}{3}\left[Q^{-1}\left(\frac{Pe_u}{4}\right)\right]^2, \tag{6.17}$$

with Q-function defined in (5.6).

All waveforms other than $(n, 1)$, that is, waveforms (n, l) with $l > 1$, need extra protection to account for the presence of the smaller signals. This is depicted in Figure 6.10 and Figure 6.11. Thus, for waveforms with $l > 1$, the distance between the constellation symbols takes the following form

$$d_{u,n,l} = d'_{u,n,l} + D_{\overline{u},n,l-1}, \tag{6.18}$$

where $D_{\overline{u},n,l-1}$ expresses the width of the $l-1$ constellation and $\overline{u} = \{u|C_{u,n,l-1} = 1\}$ indicates a user that is allocated a waveform $(n, l-1)$.

The constellation width, for the lowest $l = 1$ waveforms $(n, 1)$ can be expressed by

$$D_{u,n,1} = (\sqrt{M} - 1)d_{u,n,1}, \tag{6.19}$$

while for waveforms (n, l) with $l > 1$, the presence of underlying signals should be compensated, which is expressed by

$$D_{u,n,l} = (\sqrt{M} - 1)d_{u,n,l} + D_{\overline{u},n,l-1}. \tag{6.20}$$

6.3.2 Resource allocation problem and problem reformulation

Resource allocation problem for a single-cell OFDMA system with intra-cell reuse of subcarriers can be expressed as follows. Find such allocation arrays $\boldsymbol{C} = [C_{u,n,l}]$ and $\boldsymbol{P}_{min} = \min(\boldsymbol{P}) = [P_{u,n,l}^{min}]$ so that the traffic requirements of all users are met at the minimal cost, which is the total transmit power. This is formally written in the following relations

$$\min \sum_{u=1}^{U} \sum_{n=1}^{N} \sum_{l=1}^{L} C_{u,n,l} P_{u,n,l} \tag{6.21}$$

subject to

$$R_u \leq oR_u = \frac{b}{T_e} \sum_{n=1}^{N} \sum_{l=1}^{L} C_{u,n,l};\ u = 1, \ldots, U, \tag{6.22}$$

$$\sum_{u=1}^{U} C_{u,n,l} \leq 1;\ n = 1, \ldots, N;\ l = 1, \ldots, L, \tag{6.23}$$

$$\sum_{u=1}^{U} \sum_{l=1}^{L} C_{u,n,l} \leq L;\ n = 1, \ldots, N, \tag{6.24}$$

$$C_{u,n,l} \in \{0, 1\};\ u = 1, \ldots, U;\ n = 1, \ldots, N;\ l = 1, \ldots, L. \tag{6.25}$$

Constraint (6.22) expresses user data rate requirement R_u vs. offered data rate oR_u and indicates that a user u is allocated as many waveforms (n, l) as necessary to satisfy its data rate requirement. Constraint (6.23) indicates that a waveform (n, l) is allocated to at most one user u. This still allows for allocation of a single subcarrier to more than one user (i.e., to at most L users), which is expressed by constraint (6.24). Constraint (6.25) indicates that entries in the allocation array $[C_{u,n,l}]$ can take values $\{0\}$ or $\{1\}$. In addition, it should be noted that for $l > 1$, allocation of a waveform $(\overline{n}, l)$ involving a particular subcarrier $\overline{n}$, thus $C_{u,\overline{n},l} = 1$ for $u = 1, \ldots, U$, is only possible when a waveform $(\overline{n}, l-1)$ on that particular subcarrier $\overline{n}$ is also allocated (i.e., $C_{u,\overline{n},l-1} = 1$ for $u = 1, \ldots, U$). This follows from the previous discussion reflected in (6.13)-(6.15).

The presented allocation problem has a major difficulty, which is a mutual dependency of the entries in the cost matrix $\boldsymbol{P} = [P_{u,n,l}]$. This makes classical solutions to the assignment problem [25, 26] not possible to apply. The cost entries $P_{u,n,l}$ for waveforms (n, l) depend on the allocations made for $(n, l-1)$, see equations (6.13)-(6.20), since we have to compensate for the smaller signals already present. This can be explained by the following real-life example. Let us assume that users $u = 1, \ldots, U$ correspond to tasks, 4-QAM signals $l = 1, \ldots, L$ to workers and subcarriers $n = 1, \ldots, N$ to organization units and that $[P_{u,n,l}]$ represents the labor cost of worker l within unit n doing task u. The workers in the units are organized in a hierarchy: the labor cost of every higher-grade worker $l + 1$ consists

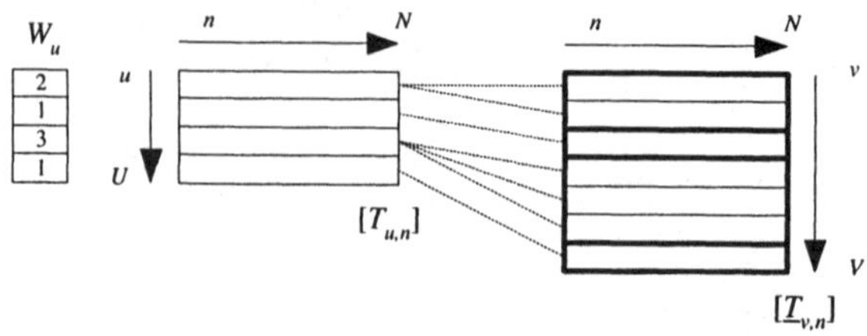

Figure 6.13 Transformation of actual users $u = 1, \ldots, U$, each requiring a data rate equivalent to W_u M-QAM symbols, to virtual users $v = 1, \ldots, V$ each requiring only one M-QAM symbol.

of his or her own labor cost, which depends on the allocated task and the supervision cost of the lower-grade worker l, which depends on his or her allocated task. It is possible that several workers l work on the same task u. Only the labor cost of worker $l = 1$, lowest in the hierarchy, does not include the cost of supervising other workers. Obviously, we are unable to determine the labor cost of worker l without knowing the tasks assigned to a supervised worker $l - 1$. The allocation problem is, of course, to find such distribution of tasks $u = 1, \ldots, U$, units $n = 1, \ldots, N$ and workers $l = 1, \ldots, L$ so that the total labor cost is minimized.

In the considered system, where an M-QAM symbol carrying $b = \log_2 M$ bits is the smallest resource to be assigned to a user, submitting the rate request R_u is equivalent to specifying the number W_u of M-QAM symbols, where $W_u = \lceil \frac{R_u T_e}{b} \rceil = \frac{o R_u T_e}{b}$. In this way, each user u requiring the data rate R_u, which is equivalent to W_u M-QAM symbols, can now be exchanged for W_u new, so-called *virtual* users, each requiring allocation of only one M-QAM symbol (i.e., a waveform indexed by (n, l)), as shown in Figure 6.13. Virtual users are indexed by v, where $v = 1, \ldots, V$ and $V = \sum_u W_u \leq N \cdot L$. This allows us to reformulate the original problem (6.21)-(6.25) into a form similar to the classical matching problem where one virtual user v is assigned one waveform (n, l) and one waveform (n, l) is assigned to at most one virtual user v. The above transformation has been introduced in the OFDMA scenario in [27,28].

Following the above transformation, the channel gain $[\underline{G}_{v,n}]$ and the SNR gap $[\underline{\Gamma}_v]$ corresponding to a virtual user v are the same as of the corresponding actual user u. Then, treating virtual users v as actual users u thus using v instead of u in relations (6.13)-(6.16), we determine the following: normalized channel gain $[\underline{T}_{v,n}]$, distance between the constellation points $[\underline{d}'_{v,n,l}]$, constellation width $[\underline{D}_{v,n,l}]$, distance between the constellation symbol $[\underline{d}_{v,n,l}]$ and finally average power $[\underline{P}_{v,n,l}]$ of a waveform (n, l) allocated virtual user v. Knowing the actual users' rate requirements, it is easy to return from the allocation of virtual users to the allocation of actual users. In the following, all variables related to virtual users will be indicated by underline, $\underline{(\cdot)}$.

Taking the above transformation into account, the original allocation problem (6.21)-(6.25) can now be reformulated. The goal is to find such allocation arrays $\underline{\boldsymbol{C}} = [\underline{C}_{v,n,l}]$ and powers $\underline{\boldsymbol{P}}_{min} = \min(\underline{\boldsymbol{P}}) = [\underline{P}^{min}_{u,n,l}]$ so that the virtual users' traffic requirements are met at the cost of minimum total transmit power. The allocation problem - reformulated can be expressed by

$$\min \sum_{v=1}^{V} \sum_{n=1}^{N} \sum_{l=1}^{L} \underline{C}_{v,n,l} \underline{P}_{v,n,l} \tag{6.26}$$

subject to

$$\sum_{n=1}^{N} \sum_{l=1}^{L} \underline{C}_{v,n,l} = 1; \; v = 1, \ldots, V, \tag{6.27}$$

$$\sum_{v=1}^{V} \underline{C}_{v,n,l} \leq 1; \; n = 1, \ldots, N; \; l = 1, \ldots, L, \tag{6.28}$$

$$\sum_{v=1}^{U} \sum_{l=1}^{L} \underline{C}_{v,n,l} \leq L; \; n = 1, \ldots, N, \tag{6.29}$$

$$\underline{C}_{v,n,l} \in \{0, 1\}; \; v = 1, \ldots, V; \; n = 1, \ldots, N; \; l = 1, \ldots, L. \tag{6.30}$$

Constraint (6.27) indicates that a virtual user is allocated only one waveform (n, l), while constraint (6.28) indicates that a waveform (n, l) is allocated to at most one virtual user v. The meaning of constraints (6.29)-(6.30) is the same as in the previous problem. Also here, for $l > 1$, allocation of a waveform $(\overline{n}, l)$, thus $C_{u,\overline{n},l} = 1$ for $u = 1, \ldots, U$, is only possible when a waveform $(\overline{n}, l-1)$ is also allocated (i.e., $C_{u,\overline{n},l-1} = 1$ for $u = 1, \ldots, U$). The applicable form of a waveform allocation array is depicted in the left-hand picture of Figure 6.14.

Another way of representing the considered allocation problem is to use variable w indicating a waveform, instead of using (n, l), where $w = l + (n-1)L$, $w = 1, \ldots, LN$ with waveform allocation array arrangement shown in the right-hand picture of Figure 6.14. Then, the allocation problem - reformulated can be expressed as

$$\min \sum_{v=1}^{V} \sum_{w=1}^{NL} \underline{C}_{v,w} \underline{P}_{v,w} \tag{6.31}$$

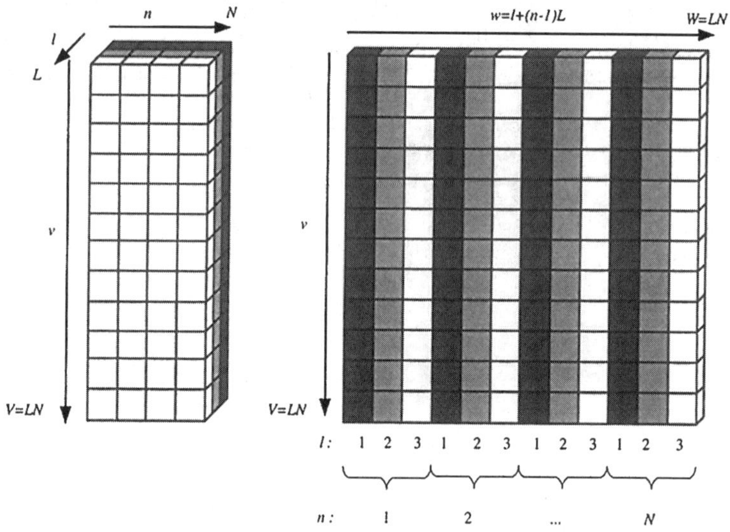

Figure 6.14 Possible ways to represent the considered waveform allocation arrays and matrices: l - layer index, n - subcarrier index, v - *virtual* user index, and w - waveform index. Different shades indicate different layers.

subject to

$$\sum_{w=1}^{NL} \underline{C}_{v,w} = 1;\ v = 1, \ldots, V, \tag{6.32}$$

$$\sum_{v=1}^{V} \underline{C}_{v,w} \leq 1;\ w = 1, \ldots, NL, \tag{6.33}$$

$$\underline{C}_{v,w} \in \{0, 1\};\ v = 1, \ldots, V;\ w = 1, \ldots, NL. \tag{6.34}$$

The allocation problems (6.26)-(6.30) and (6.31)-(6.34) resemble the 2-dimensional matching problem [26], in which one dimension corresponds to users v and one to waveforms indicated as w or as (n, l). However, also here, the mutual dependency of the entries in the cost matrix is present, which makes this not a classical matching problem and which constitutes a major difficulty in solving it.

6.4 RRM ALGORITHMS FOR DOWNLINK OFDMA SYSTEMS WITH INTRA-CELL SUBCARRIERS REUSE

In this subsection, a set of algorithms is described that solve the RRM problems (6.21)-(6.25) and (6.26)-(6.30) of the single-cell downlink OFDMA system with intra-cell reuse of subcarriers enabled by the application of embedded modulation.

In the design of the algorithms, the general approach introduced in Section 5.3 is applied: an initial allocation resulting in a fast, rough but good solution, is followed by an iterative swapping of chosen entries in order to improve the initial allocation if the time permits. Such time-scalability is especially applicable to adaptive communication systems, since the algorithm may stop its operation depending on the available time. The following algorithms are proposed for the initial allocation phase: *basic dough laying*, *improved dough laying*, and *one shot*. Each of them may be followed by the *embedded swapping* routine in order to improve the initial performance. For all the proposed algorithms, the transformation from actual users $u = 1, \ldots, U$ to *virtual* users $v = 1, \ldots, V$ is adopted.

6.4.1 Dough laying algorithm: basic and improved

In this subsection we describe two algorithms for the initial resource allocation phase in the downlink OFDMA system with intra-cell reuse of subcarriers, namely, *basic dough laying* and *improved dough laying*. Because both routines are closely

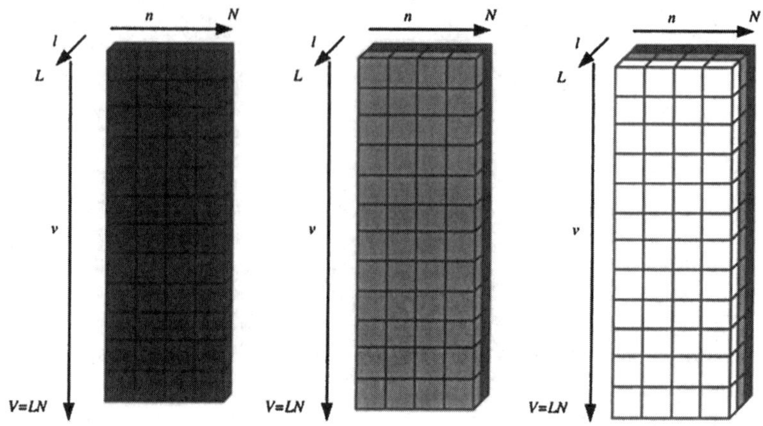

Figure 6.15 The basic principle of the dough laying algorithm: distribute N subcarriers at layer l among users experiencing better channel conditions before distributing N subcarriers at layer $l+1$.

related, first a common part of both is presented and after that the differences between the two algorithms are identified.

Let us define a layer l as all the N waveforms $(n, l), n = 1, \ldots, N$ with index l, as shown in Figure 6.14. In the *dough laying* algorithm waveforms are allocated to users in a "layer-by-layer" fashion. This is depicted in Figure 6.15, where in L steps (here $L = 3$) all the waveforms are distributed. At each layer l, allocation of waveforms boils down to allocation of N subcarriers among users experiencing best channel conditions before going to layer $l+1$ where subcarriers are distributed among users with worse channels. Thanks to such an approach, the problematic interdependency of the entries in the cost matrix, (6.21)-(6.26) is overcome by having determined the constellation design parameters for waveforms belonging to layer l before processing allocation in layer $l+1$.

The fundamental metric used for making allocations at each layer l is the normalized channel gain matrix $\Upsilon_{v,n,l}, v = 1, 2, \ldots, V_l$, where $V_l = V - (l-1)N$ is the number of virtual users at layer l with no waveform allocated where $V = \sum_u W_u$. For layer $l = 1$, the normalized channel gain matrix is $\Upsilon_{v,n,1} = \underline{T}_{v,n}$ as defined in (6.16) with v used instead of u. For $l > 1$, the normalized channel gain matrix $\Upsilon_{v,n,l}$ is formed from the previous matrix $\Upsilon_{v,n,l-1}$ as depicted in Figure 6.16. Rows corresponding to virtual users $\{\overline{v}_n\}, n = 1, \ldots, N$ already allocated in

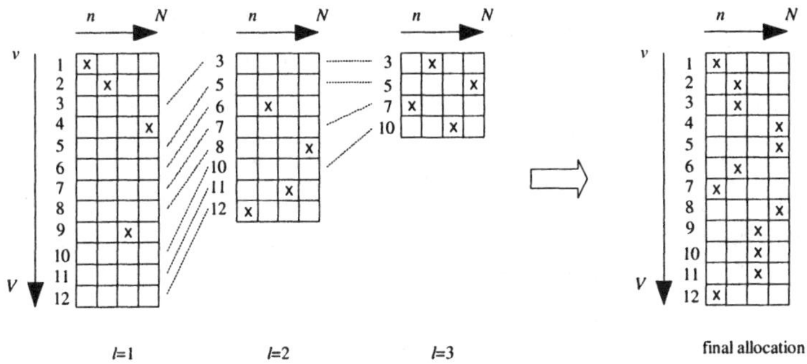

Figure 6.16 Transformations of the normalized channel gain matrix $\Upsilon_{v,n,l}$ in the *dough laying* algorithm. Allocations of subcarriers n to users v are indicated by "x."

layer $(l-1)$ are removed from $\Upsilon_{v,n,l-1}$. The resulting matrix is $\Upsilon_{v,n,l}$. In this way $\Upsilon_{v,n,l}$ is built out of channel conditions corresponding only to those virtual users that do not yet have allocated a waveform, while knowing the constellation design parameters of the already made allocations in lower layers. In the next step, the *dough laying* algorithm determines the embedded constellation design parameters, such as $\underline{d}'_{v,n,l}$, $\underline{d}_{v,n,l}$, and $\underline{D}_{v,n,l}$ according to (6.18)-(6.16), using Υ instead of T and v instead of u. Knowing the constellation design parameters we are able to determine the cost matrix $[\underline{P}_{v,n,l}]$ as defined in (6.13). Then, we can state a new allocation problem for particular layer l as follows

$$\min \sum_{v=1}^{V_l} \sum_{n=1}^{N} \underline{C}_{v,n,l} \underline{P}_{v,n,l} \tag{6.35}$$

subject to

$$\sum_{v=1}^{V_l} \underline{C}_{v,n,l} = 1; \; n = 1, \ldots, N, \tag{6.36}$$

$$\sum_{n=1}^{N} \underline{C}_{v,n,l} \leq 1; \; v = 1, \ldots, V_l, \tag{6.37}$$

$$\underline{C}_{v,n,l} \in \{0,1\}, \; v = 1, \ldots, V_l; \; n = 1, \ldots, N. \tag{6.38}$$

In this way, the original allocation problem of (6.26)-(6.30) is transformed into L 2-dimensional allocation problems, each for one layer l. Since layer index l in

(6.35)-(6.38) is not an optimization variable, the problem (6.35)-(6.38) is a classical matching problem, where at each layer l we need to distribute N waveforms among N virtual users. Since we aim at minimization of the total transmit power, where first users with better channel conditions are processed before users with worse channel conditions, such a hierarchical allocation forms hierarchical, embedded constellation on each subcarrier, as shown in Figure 6.10 and Figure 6.11. In the embedded constellation the waveforms corresponding to better channel conditions are centered around the waveforms corresponding to worse channel conditions.

Algorithm 6.1: Basic dough laying routine for a single-cell OFDMA system with intra-cell reuse of subcarriers.

---

input : $[\underline{T}_{v,n}]$
output: $[\underline{C}_{v,n,l}]$ and $[\underline{P}^{min}_{v,n,l}]$

begin
  for $l = 1$ **to** L **do**
    { Determine the normalized channel gain matrix $\{\Upsilon_{v,n,l}\}$: }
    $\Upsilon_{v,n,l} = \underline{T}_{v,n}$ using (6.16) for $l = 1$ or
    $\Upsilon_{v,n,l} = \Upsilon_{v,n,l-1} - \{\overline{v}_n,\ n = 1, \ldots, N\}$ for $l > 1$
    Calculate the constellation design parameters: $\underline{d}'_{v,n,l}$, $\underline{d}_{v,n,l}$ and $\underline{D}_{v,n,l}$ using (6.18)-(6.16) with Υ instead of T and v instead of u
    Determine the cost matrix $[\underline{P}_{v,n,l}]$ using (6.13)
    {Find the initial waveform allocation matrix $[\underline{C}_{v,n,l}]$ at layer l: }
    for $v = 1$ **to** V_l **do**
      $(\overline{n}, l) = \arg\min_n \underline{P}_{v,n,l}$
      if $\underline{C}_{v,\overline{n},l} = 0$ **then**
        $\underline{C}_{v,\overline{n},l} = 1$
    Find corresponding power allocation $[\underline{P}^{min}_{v,n,l}]$ at layer l
    If improved dough laying then apply Algorithm 6.2
end

The problem (6.35)-(6.38) can be solved optimally with the use of the Hungarian algorithm [25, 26]. However, aiming at a low complexity solution a simple suboptimal routine based on greedy selection of the best subcarriers by users is applied. The routine is based on Algorithm 5.4. The difference here is the search for the cheapest waveforms in terms of transmit power instead of the search for

Algorithm 6.2: Iterative swapping subroutine needed for improved dough laying for single-cell OFDMA system with intra-cell reuse of subcarriers.

input : $[\underline{T}_{v,n}]$, l, initial $[\underline{C}_{v,n,l}]$ and $[\underline{P}^{min}_{v,n,l}]$
output: improved $[\underline{C}_{v,n,l}]$ and $[\underline{P}^{min}_{v,n,l}]$

begin
  { For a given layer l }
  repeat
    for *each pair of users* $(v_1, v_2) \in \{1, \ldots, V_l\}$ **do**
      read original waveform indices: $(n_1, l) \rightarrow v_1; (n_2, l) \rightarrow v_2$
      if $(n_1, l) = \oslash$ **then**
        swap allocated waveform: $(n_2, l) \rightarrow v_1$
        calculate $\Delta P = P_{new} - P_{old}$ where $P_{new} = \underline{P}_{v_1,n_2,l}$ and $P_{old} = \underline{P}_{v_2,n_2,l}$

      else if $(n_2, l) = \oslash$ **then**
        swap allocated waveform: $(n_1, l) \rightarrow v_2$
        calculate $\Delta P = P_{new} - P_{old}$ where $P_{new} = \underline{P}_{v_2,n_1,l}$ and $P_{old} = \underline{P}_{v_1,n_2,l}$
      else
        swap allocated waveforms: $(n_1, l) \rightarrow v_2$ and $(n_2, l) \rightarrow v_1$
        calculate $\Delta P = P_{new} - P_{old}$ where $P_{new} = \underline{P}_{v_1,n_1,l} + \underline{P}_{v_2,n_2,l}$ and $P_{old} = \underline{P}_{v_2,n_1,l} + \underline{P}_{v_1,n_2,l}$

    $(\overline{v}_1, \overline{v}_2, \overline{n}_1, \overline{n}_2) = arg\,\max_{v,n} \Delta P$
    if $\max \Delta P > 0$ **then**
      {apply swapping in $\underline{C}_{v,n,l}$: }
      $(\overline{n}_1, l) \rightarrow \overline{v}_2; (\overline{n}_2, l) \rightarrow \overline{v}_2$
      update $[\underline{C}_{v,n,l}]$ and $[\underline{P}^{min}_{v,n,l}]$
  until $\Delta P > 0$
end

the best quality subcarriers. The routine is triggered for each layer l, starting from $l = 1$ and ending at $l = L$. Also here, at each layer l, the initial allocation may be further improved by the application of iterative swapping. For this purpose, we use a modification of Algorithm 5.5. The difference here is the need for the ability to handle waveforms that have not yet been allocated. The dough laying algorithm implementing only the initial allocation at each layer is referred to as basic dough laying. If initial allocation at each layer is followed by the iterative swapping at each layer, then the resource allocation algorithm is referred to as improved dough laying.

We present the basic dough laying routine as Algorithm 6.1, which after application of *iterative swapping* presented in Algorithm 6.2 becomes the improved dough laying.

Both the basic dough laying and the improved dough laying algorithms can be further improved by the application of the embedded swapping algorithm, which is described further in this chapter.

6.4.2 One-shot algorithm

In the basic and the improved dough laying algorithms, the problematic dependencies of the cost entries for waveforms (n, l) on the allocations made for $(n, l-1)$, (6.26)-(6.30), are overcome by allocating waveforms in a layer-by-layer fashion, where at each layer l, the problem (6.35)-(6.38) is solved. The idea behind the alternative one-shot algorithm is to substitute the exact values of the $l-1$ layer's constellation widths $\underline{D}_{v,n,l-1}$ by the values averaged over V users and then to distribute all the waveforms among all the users in one move (or one-shot). The average widths over V users are defined as

$$\widetilde{\underline{D}}_{n,l-1} = \frac{1}{V}\sum_{v=1}^{V} \underline{D}_{v,n,l-1}, \tag{6.39}$$

where $l = 2, \ldots, L; n = 1, \ldots, N$. Using (6.39) in (6.20) allows to compose the cost matrix $[\widetilde{\underline{P}}_{v,n,l}]$, according to (6.13), where $\widetilde{(\cdot)}$ is used to indicate averaging modification applied to $\widetilde{\underline{D}}_{n,l-1}$. Although the cost matrix created in such a way does not contain the exact values of the transmit power, it is free from the problematic interdependencies. Then, the allocation problem becomes a classical 2-dimensional matching problem, where the cost matrix does not have the problematic recursion. To reflect it better we use w to indicate a waveform instead of (n, l), where $w = l + (n-1)L$, $w = 1, \ldots, LN$. The allocation problem for the one-shot algorithm can be expressed as

$$\min \sum_{v=1}^{V} \sum_{w=1}^{NL} \underline{C}_{v,w} \underline{\widetilde{P}}_{v,w} \tag{6.40}$$

subject to

$$\sum_{w=1}^{NL} \underline{C}_{v,w} = 1; \; v = 1, \ldots, V, \tag{6.41}$$

$$\sum_{v=1}^{V} \underline{C}_{v,w} \leq 1; \; w = 1, \ldots, NL, \tag{6.42}$$

$$\underline{C}_{v,w} \in \{0, 1\}; \; v = 1, \ldots, V; \; w = 1, \ldots, NL. \tag{6.43}$$

The above problem, as a classical 2-dimensional matching problem, can be solved optimally using routines [25, 26], or suboptimally using Algorithms 5.4 and 5.5. After solving this problem, based on the resulting waveform allocation matrix $[\underline{C}_{v,n,l}]$, the transmit power $[\underline{P}^{min}_{v,n,l}]$ needs to be calculated using the exact constellation design parameters (and not the average values as in (6.39)) in order to remove the averaging effect.

6.4.3 Embedded swapping algorithm

To further improve the initial allocation obtained by one of the previously presented schemes, including basic dough laying, improved dough laying, or one shot, an additional iterative swapping algorithm is described in this subsection. This new algorithm is meant to reallocate waveforms belonging to any layer of each subcarrier's embedded constellation, which is in contrast to the improved dough laying algorithm, where reallocations are limited to only one layer l. Therefore it is referred to as the embedded swapping algorithm. Its operation is described in the following.

At each iteration step, all possible user pairs $(v_1, v_2) \in \{1, \ldots, V\}$ are considered and for each pair the prospective power reduction resulting from swapping waveforms between those users is evaluated. In the first instance, the algorithm determines the indices of waveform (n_1, l_1) originally allocated to user v_1 (i.e., where $\underline{C}_{v_1,n_1,l_1} = 1$) and of waveform (n_2, l_2) originally allocated to user v_2 (i.e., where $\underline{C}_{v_2,n_2,l_2} = 1$). Then reallocation is performed by swapping waveforms between users (v_1, v_2). As a result, waveform (n_1, l_1) is reallocated to user v_2 (i.e., $\underline{C}_{v_2,n_1,l_1} = 1$) and waveform (n_2, l_2) is reallocated to user v_1 (i.e., $\underline{C}_{v_1,n_2,l_2} = 1$),

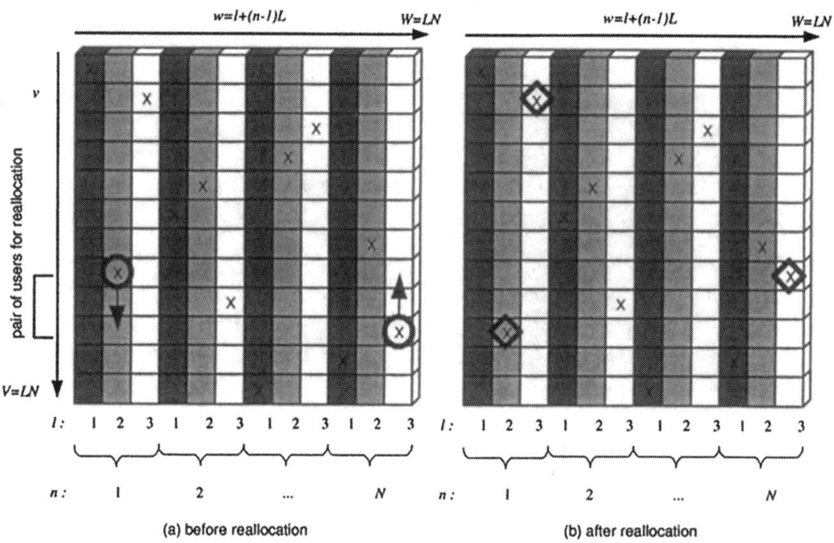

Figure 6.17 Swapping of waveforms between two users in downlink OFDMA system employing embedded modulation to reuse subcarriers within a cell. Circle indicates entries to be reallocated, while diamond indicates allocated waveforms that need to be used in calculating power reduction. Please note that not only the involved waveform allocations are taken into account but also all the allocated waveforms that are located above the involved ones in a layer hierarchy.

Algorithm 6.3: Embedded swapping routine for a single-cell OFDMA system with intra-cell reuse of subcarriers.

---

input : initial $[\underline{C}_{v,n,l}]$
output: improved $[\underline{C}_{v,n,l}]$ and $[\underline{P}^{min}_{v,n,l}]$

begin
    repeat
        for *each pair of users* $(v_1, v_2) \in \{1, \ldots, V\}$ **do**
            {read original waveform indices: }
            $(n_1, k_1) \rightarrow v_1; (n_2, l_2) \rightarrow v_2$
            {swap allocated waveforms: }
            $(n_1, l_1) \rightarrow v_2; (n_2, l_2) \rightarrow v_1$
            calculate $\Delta P = P_{old} - P_{new}$ using (6.45), (6.46)
        $(\overline{v_1}, \overline{v_2}, \overline{n_1}, \overline{n_2}, \overline{l_1}, \overline{l_2}) = \arg\max_{v,n,l} \Delta P$
        if $\max \Delta P > 0$ **then**
            {apply swapping in $[\underline{C}_{v,n,l}]$: }
            $(\overline{n_1}, \overline{l_1}) \rightarrow \overline{v_2}; (\overline{n_2}, \overline{l_2}) \rightarrow \overline{v_1}$
        update power allocation $[\underline{P}^{min}_{v,n,l}]$
    until $\Delta P > 0$
end

---

as can be seen in the example shown in Figure 6.17. For the new swapped allocation the algorithm computes the prospective power reduction factor as

$$\Delta P = P_{old} - P_{new}, \tag{6.44}$$

where

$$P_{old} = \sum_{l=l_1}^{L} \sum_{v=1}^{V} (\underline{P}_{v,n_1,l} | \underline{C}_{v_1,n_1,l_1} = 1) + \sum_{l=l_2}^{L} \sum_{v=1}^{V} (\underline{P}_{v,n_2,l} | \underline{C}_{v_2,n_2,l_2} = 1), \tag{6.45}$$

$$P_{new} = \sum_{l=l_1}^{L} \sum_{v=1}^{V} (\underline{P}_{v,n_1,l} | \underline{C}_{v_2,n_1,l_1} = 1) + \sum_{l=l_2}^{L} \sum_{v=1}^{V} (\underline{P}_{v,n_2,l} | \underline{C}_{v_1,n_2,l_2} = 1). \tag{6.46}$$

It can be seen that in determining the prospective power reduction, not only the involved waveforms (n_1, l_1) and (n_2, l_2) are involved, but also all the waveforms of the higher layers, $l \geq l_1$ and $l \geq l_2$. This is due to hierarchical signals' stacking, where a change done at a certain layer propagates over all layers above. This is also reflected in Figure 6.17 (b), where the allocated waveforms that should be taken into account in power reduction calculation are indicated by a diamond. In the next step, the algorithm picks the maximal cost reduction factor. If it is positive then the decision is made to reallocate waveforms between users (v_1, v_2), so that $(n_1, l_1) \rightarrow v_2$; $(n_2, l_2) \rightarrow v_1$. If it is not positive then the old waveform allocation remains, since there is no point in reallocating waveforms between users of this users' pair. Finally, the waveform and power allocation arrays are updated and the algorithm proceeds to the next iteration step (i.e., a new users' pair). Also this iterative algorithm is time-scalable, which means that its operation may be stopped earlier if a lower performance is acceptable and/or the available time has elapsed. The complete embedded swapping algorithm is outlined as Algorithm 6.3.

6.5 NUMERICAL EXPERIMENTS

In this section we evaluate by numerical experiments the performance of the described algorithms for the single-cell downlink OFDMA system with intra-cell reuse of subcarriers. For a simulation scenario we apply the channel model as presented in Chapter 2 and the traffic model as presented in Chapter 5. Unless stated otherwise, main system parameters are as introduced in Chapter 4, including system bandwidth BW = 7 MHz, mean rms delay spread τ_{rms} = 264 ns, path loss exponent $\alpha = 4.4$, mean Rician $K_{factor} = -3$ dB , attenuation $PL(d_o) = -39$ dB at the reference distance of $d_o = 1$ m, and shadowing variance $\sigma^2_{shadow} = 18$ dB. For simplicity, each user is assumed to require the same bit rate $R_u = R$ and the same bit error rate $Pe_u = Pe$.

With these system parameters, the performance of the described algorithms have been compared. In addition, their performance compared to other RRM schemes has been investigated under the condition of changing the following system / environmental parameters:

- Total system load $R_{total} = \{\frac{1}{3}R^{max}_{total}, \frac{2}{3}R^{max}_{total}, R^{max}_{total}\}$, where $R_{total} = \sum_u R_u$ is the sum traffic in [bit/s] required by all users and $R^{max}_{total} = \frac{N \cdot L \cdot b}{T_e}$ is the maximal traffic in [bit/s] that can be offered by the system;
- Number of users U in a fully-loaded system (i.e., when $R_{total} = R^{max}_{total}$);
- Mean rms delay spread $\tau_{rms} = \{50, 200, 400, 800\}$ ns;
- Path loss exponent $\alpha = \{2.5, 3, 3.5, 4, 4.5, 5\}$;
- Variance due to shadowing $\sigma^2_{shadow} = \{10, 14, 18, 22\}$ dB.

6.5.1 Example allocation patterns

Example allocation patterns obtained by the use of the improved dough laying algorithm followed by embedded swapping, for $N = 16$ and $L = 3$, for a fully loaded system, are shown in Figure 6.18. It can be seen that waveforms on the same subcarrier may be allocated to different users or to the same user, depending on the users' needs and channel conditions. Waveforms tend to be allocated to users in a way that good channel conditions are preferred. In the lower picture of Figure 6.18 depicting transmit power allocation, the layered hierarchy of the allocations is clearly visible, where layer $l = 1$ is the best one with the least transmit power required, while $l = 3$ is the worst one with the highest transmit power required. Since we have $L = 3$ layers and all are filled with traffic, the embedded constellation has a form of 64-QAM as shown in Figure 6.11. If the total traffic demand is lower than maximal system capacity or if we have less layers in the system, the embedded constellation will takc a form of 16-QAM as shown in Figure 6.10, or of 4-QAM.

Another representation of the same case is depicted in Figure 6.19, where bit loading is shown explicitly by indicating the number of bits b offered to a particular user. Also here, it is visible that users reuse the same subcarrier within one cell. Moreover, a subcarrier may carry bits addressed to different users, since a

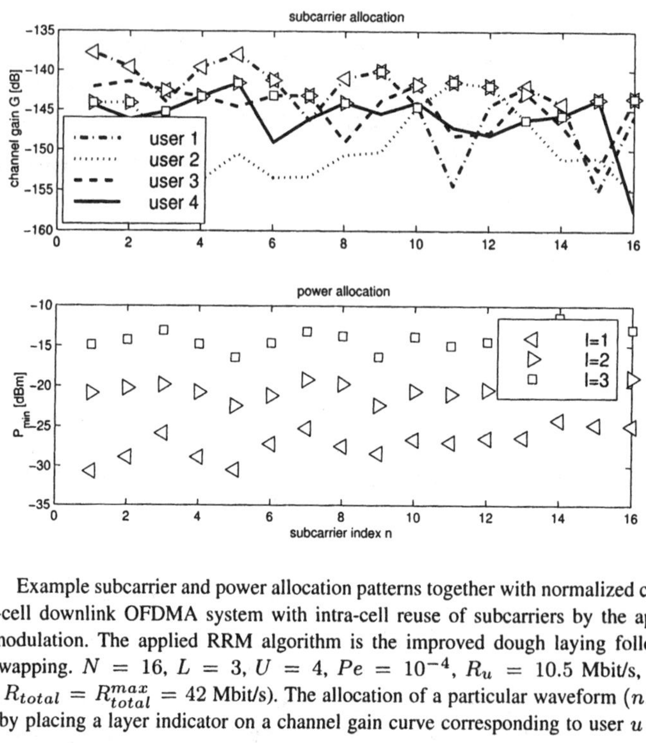

Figure 6.18 Example subcarrier and power allocation patterns together with normalized channel gains for a single-cell downlink OFDMA system with intra-cell reuse of subcarriers by the application of embedded modulation. The applied RRM algorithm is the improved dough laying followed by the embedded swapping. $N = 16$, $L = 3$, $U = 4$, $Pe = 10^{-4}$, $R_u = 10.5$ Mbit/s, fully loaded system (i.e., $R_{total} = R_{total}^{max} = 42$ Mbit/s). The allocation of a particular waveform (n, l) to user u is indicated by placing a layer indicator on a channel gain curve corresponding to user u in the upper picture.

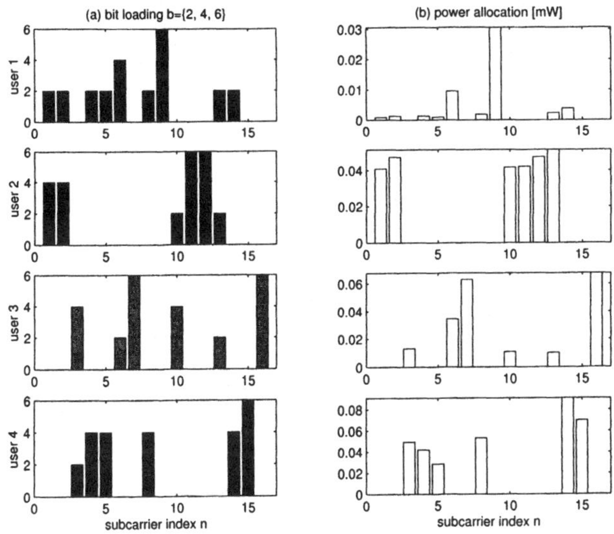

Figure 6.19 Another representation of bit and power allocation patterns (corresponding to Figure 6.18).

subcarrier may be allocated to more than one user within the same cell. This intra-cell reuse of subcarriers by the application of embedded modulation may be treated as a generalization of adaptive modulation, where bits loaded on one subcarrier are not restricted to one user only. As such, the final embedded 64-QAM constellation can be built out of component 4-QAM constellations addressed to different users, while a classical 64-QAM constellation can only be addressed to one user. This extra flexibility allows for a better signal matching and higher power efficiency.

For comparison, bit and power allocation patterns for OFDMA schemes with exclusive assignment of subcarriers to users are presented in Figure 6.20 assuming the same channel and traffic conditions as in Figure 6.18 and Figure 6.19. It can be seen that due to the restriction of OFDMA exclusive to allocate subcarriers exclusively, more transmit power is necessary than in case of OFDMA embedded.

Another advantage of OFDMA with intra-cell subcarrier reuse is its ability to offer additional separable links that can be used to serve a higher number of users when compared to OFDMA with exclusive subcarrier allocation. In the case of the considered system, where $N = 16$, $L = 3$, the system can serve up to $V = N \cdot L = 48$ users (each allocated one waveform with $b = 2$ bits) thanks to the possibility to share a single subcarrier. This is L times more when compared

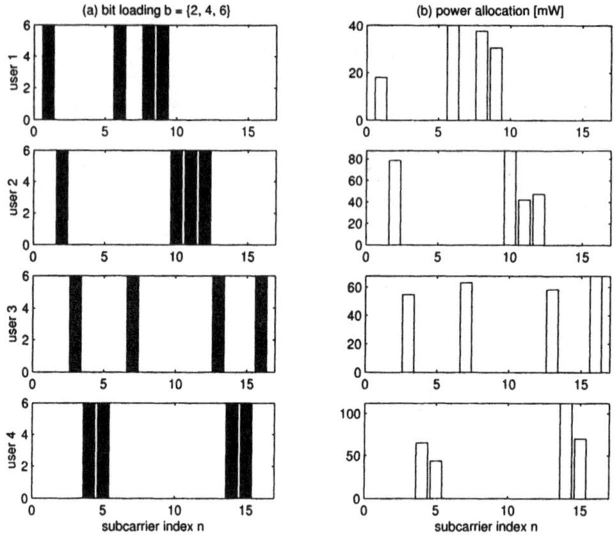

Figure 6.20 Example bit and power allocation patterns for a single-cell downlink OFDMA system with exclusive allocation of subcarriers to users. The applied RRM algorithm is the OFDMA exclusive introduced in Chapter 5. The same channel and traffic conditions are applied as in Figure 6.18.

to a system with only one signal present on each subcarrier (e.g., OFDMA with exclusive allocation). In case of such systems, a higher number of users would require distinguishing them in another domain (e.g., in time domain by the use of TDMA).

6.5.2 Convergence verification

Table 6.1

Number of Iterations Required for Embedded Swapping Algorithm 6.3 to Achieve Convergence

	Basic dough laying	*Improved dough laying*	*One-shot*
$N = 4, L = 2$	3	1	1
$N = 4, L = 4$	8	2	9
$N = 16, L = 2$	20	6	22
$N = 16, L = 4$	58	10	44

Table 6.2

Number of Iterations Required for Iterative Swapping Subroutine (Algorithm 6.2) Operating in Layer l to Achieve Convergence

	$l = 1$	$l = 2$	$l = 3$	$l = 4$
$N = 4, L = 2$	2	1		
$N = 4, L = 4$	2	2	1	1
$N = 16, L = 2$	12	8		
$N = 16, L = 4$	14	16	14	9

As indicated in Section 6.4, the algorithms used for initial resource allocation, namely, basic dough laying, improved dough laying, and one-shot, may be followed by the embedded swapping algorithm. This subsection verifies the convergence of the embedded swapping applied after one of the initial allocation algorithms. The convergence is understood as a situation when waveform reallocation does not lead to further total cost reduction. The results, averaged over 100 different channel and traffic realizations, are depicted in Figure 6.21. It can be seen that the embedded swapping algorithm converges after several iteration steps. The more layers L or more subcarriers N we have, the more iteration steps are required

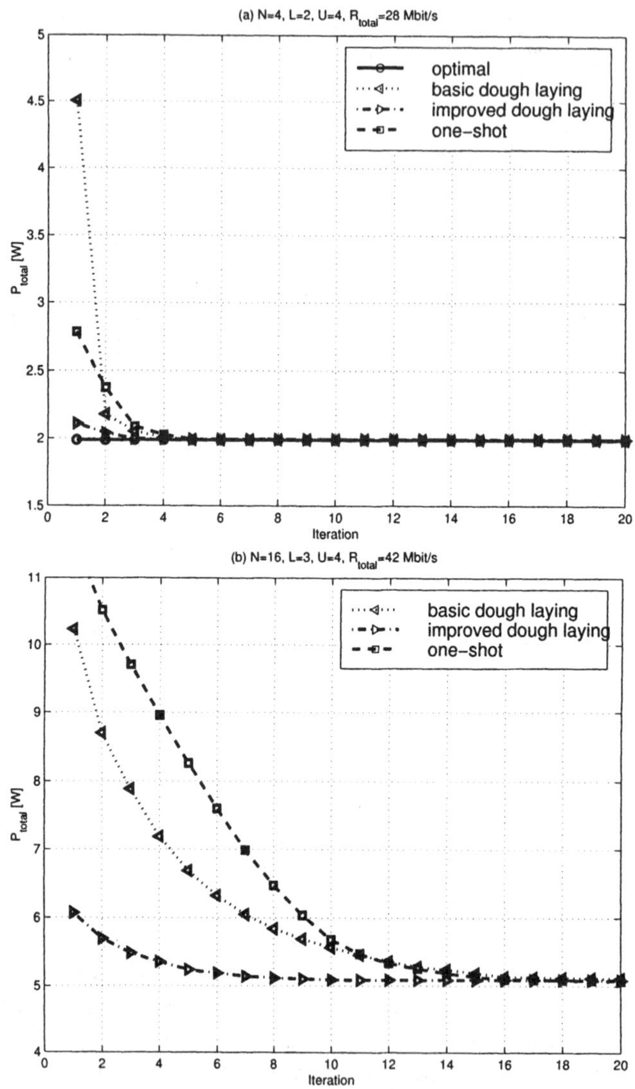

Figure 6.21 Convergence of the embedded swapping algorithm after application of the following initial allocation algorithms: basic dough laying, improved dough laying, or one-shot, in a fully loaded system.

to achieve the convergence. However, as already indicated, iterative improvement allows for performance scaling, so when the available time is short the algorithm may be stopped at the cost of a slightly lower performance. The lowest number of iteration steps required to achieve convergence is observed in case of the improved dough laying algorithm used as the initial RRM scheme. Also this algorithm exhibits the best performance (least transmit power required) at the start of the embedded swapping operation. The price for it is the increased computational complexity, since improved dough laying applies iterative swapping of subcarriers internally, at each layer. The performance of the other two algorithms expressed as the convergence time or the required transmit power is comparable.

Sometimes the performance of the suboptimal algorithm gets very close to the performance of the optimal lower-bound, but never reaches it. This is caused by the fact that reallocation of only two entries are allowed in the embedded swapping algorithm, which sometimes may not lower the overall cost, while a reallocation of more entries could do so. Thus, reallocations between all possible triples of users, quadruples, and so forth, should be considered. With this reasoning, however, we arrive at a system, where reallocation between any possible combinations of entries is possible, which is basically the exhaustive search approach. Obviously, the price to be paid for such an approach is computation time and complexity.

Table 6.1 presents the numbers (averaged over 100 channel / traffic realizations) of iterations required for the embedded swapping routine to achieve convergence. It is confirmed that the improved dough laying algorithm requires least iteration steps to achieve near-optimal performance. However, in this scheme, iterative waveform swapping is performed, after initial allocation, at every layer, and these internal reallocations are simpler (involve less calculations) than reallocations in the final embedded swapping routine. The number of iterations required to achieve convergence by the iterative swapping Algorithm 6.2 at each later l is shown in Table 6.2. Also here, the more subcarriers or layers we have, the more iteration steps are required to achieve convergence.

6.5.3 Performance comparison of the RRM algorithms

The described algorithms for RRM in a single-cell downlink OFDMA system with intra-cell reuse of subcarriers are compared based on the cumulative distribution functions of the total transmit power. The comparison is made for the basic dough laying, improved dough laying, and one-shot algorithms before and after application of the embedded swapping algorithm. The results, averaged over 100 different channel and traffic realizations, are depicted in Figure 6.22. It can be confirmed

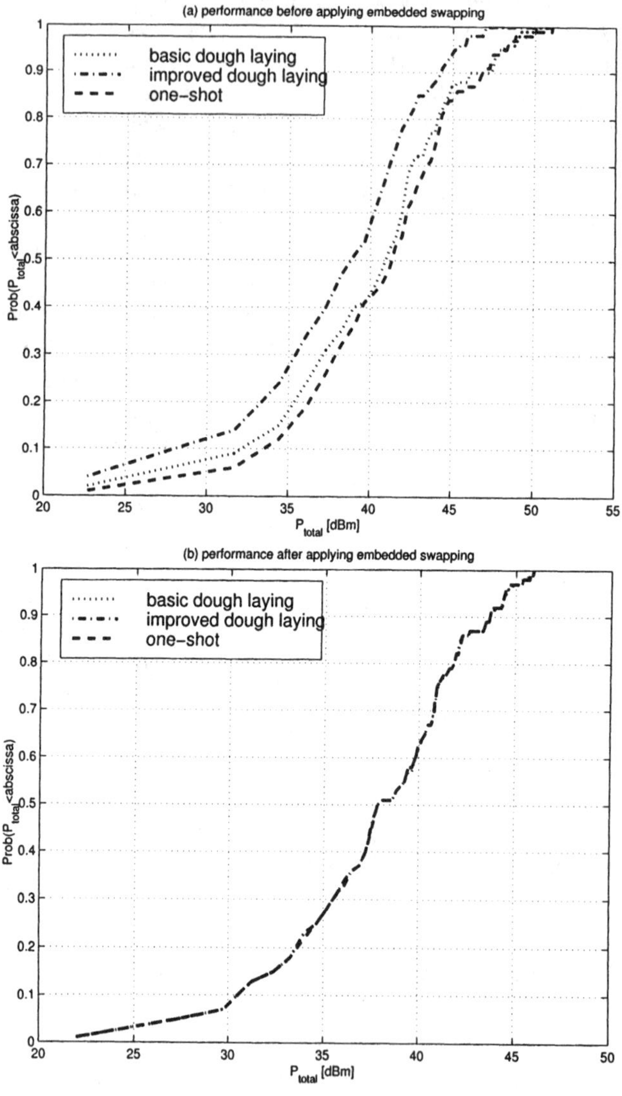

Figure 6.22 Comparison of the cumulative distribution functions of the total transmit power for a single-cell downlink OFDMA system with intra-cell reuse of subcarriers. $N = 16$, $L = 3$, $U = 4$, $Pe = 10^{-4}$ and a system fully loaded, $R_{total} = R_{total}^{max} = 42$ Mbit/s.

that the best performance before applying embedded swapping is achieved with the improved dough laying scheme, Figure 6.22 (a). The basic dough laying and the one-shot methods offer the worst performance, similar for both schemes. Also it can be confirmed that the performance of all initial schemes after applying embedded swapping is the same, as depicted in Figure 6.22 (b).

For further comparisons and evaluations, improved dough laying followed by embedded swapping is chosen as a representative scheme for a downlink OFDMA system with intra-cell reuse of subcarrier, and is referred to as the OFDMA embedded scheme.

6.5.4 Performance comparison with reference schemes

The following reference schemes are used for further evaluation of the OFDMA embedded method:

- *Lower bound (optimal)* scheme, which is basically the exhaustive search routine performed over all possible assignments limited to a case where $L = 2$ and $N = 4$ due to the excessive computational time;
- OFDMA exclusive scheme, with application of adaptive modulation. This method has been introduced in Chapter 5, where subcarriers are exclusively allocated to users (thus no subcarriers' reuse is applied);
- FDMA scheme, which is based on a conventional FDMA approach where subcarriers are fixed to users and fixed modulation level is applied on each subcarrier. Only the transmit power is adapted according to the observed channel conditions.

Again, assuming a fully loaded system, cumulative distribution functions of the total transmit power have been compared for each of the considered schemes. The results, averaged over 100 different channel and traffic realizations are depicted in Figure 6.23. It can be observed that the performance of the OFDMA embedded algorithm is very close to the lower bound (optimal) solution. Moreover, the OFDMA embedded algorithm outperforms the OFDMA exclusive method showing that it is more power-efficient to reuse subcarriers within a cell than to apply exclusive subcarrier assignment even with adaptive modulation. This is because in the case of a subcarrier reuse bits loaded on a given subcarrier are not necessarily addressed to the same user. It allows to better exploit the channel variability and thus is more power-efficient than bit loading where bits must be addressed to the same user, as is the case of classical adaptive modulation. The performance of the FDMA method is the worst because it neither gains from subcarrier reuse nor applies

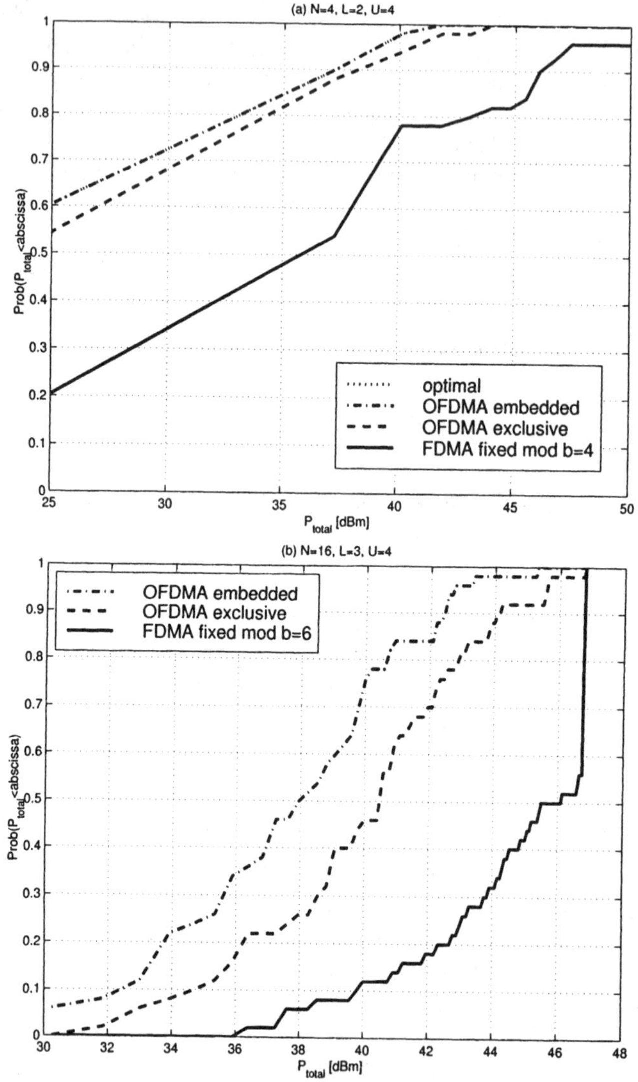

Figure 6.23 Comparison of the cumulative distribution function of the total transmit power for a single-cell downlink OFDMA system where various RRM algorithms are applied. It is assumed that a system is fully loaded. The system parameters are as follows: (a) $U = 4$, $N = 4$, $L = 2$, $Pe = 10^{-4}$, $R_{total} = R_{total}^{max} = 28$ Mbit/s; (b) $U = 4$, $N = 16$, $L = 3$, $Pe = 10^{-4}$, $R_{total} = R_{total}^{max} = 42$ Mbit/s.

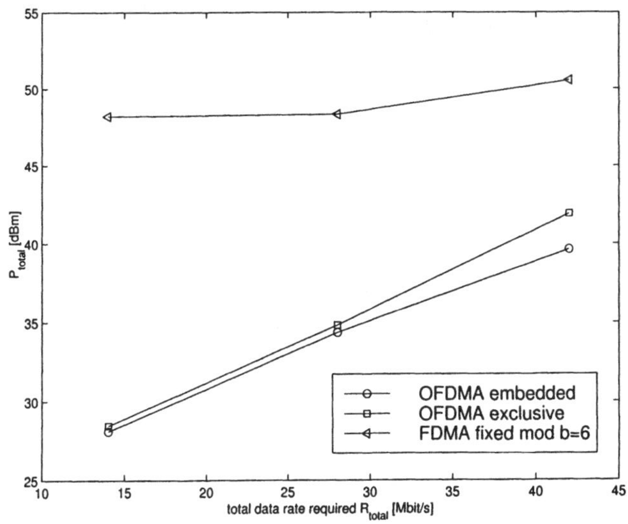

Figure 6.24 Influence of the total required traffic load R_{total}. The other system parameters are $N = 16$, $U = 4$, $\tau_{rms} = 264$ ns, $\alpha = 4.4$, σ^2_{shadow} = 18 dB, $Pe = 10^{-4}$.

adaptive subcarrier allocation. The following subsections evaluate the performance of the OFDMA embedded algorithm under various environmental conditions and compare it to the considered reference methods.

6.5.5 Influence of total system load

The performance comparison of OFDMA embedded and the considered reference schemes for various total required traffic loads $R_{total} = \{\frac{1}{3}R^{max}_{total}, \frac{2}{3}R^{max}_{total}, R^{max}_{total}\}$, where $R^{max}_{total} = 42$ Mbit/s, is depicted in Figure 6.24. It can be observed that the largest gain of OFDMA embedded over OFDMA exclusive is experienced for a fully loaded system ($R_{total} = R^{max}_{total}$). This can be explained by the fact that intra-cell reuse of subcarriers is fully exploited in this case, since a fully loaded system corresponds to all L layers loaded with traffic. For lower system loads (e.g., $\frac{1}{3}R^{max}_{total}$, which corresponds to only one layer loaded with traffic), the gain of OFDMA embedded over OFDMA exclusive is negligible since, in this case, both schemes distribute resources in almost the same way. The worst performance is observed for the FDMA scheme.

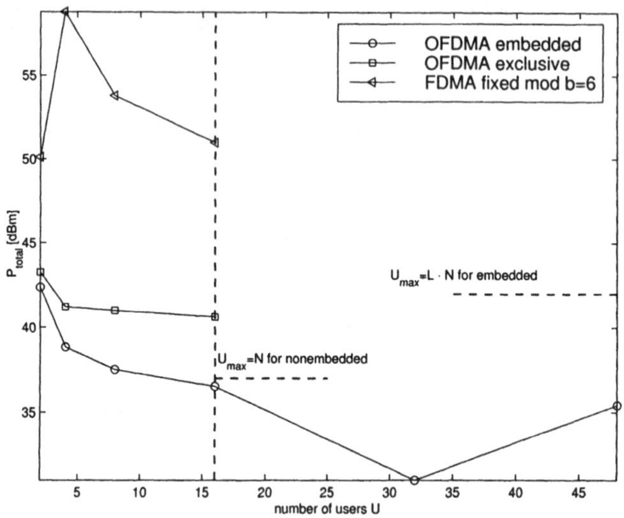

Figure 6.25 Influence of the number of users U requesting a service, assuming a fully loaded system, so $R_u = R_{total}^{max}/U$. The other system parameters are $N = 16$, $\tau_{rms} = 264$ ns, $\alpha = 4.4$, $\sigma_{shadow}^2 =$ 18 dB, $Pe = 10^{-4}$.

6.5.6 Influence of the number of users

The performance comparison of OFDMA embedded and the considered reference schemes for various numbers of users $U = \{2, 4, 8, 16, 32, 48\}$ requesting a service and assuming a fully loaded system with $N = 16$ subcarriers, is depicted in Figure 6.25. $U_{max} = N = 16$ is the maximal number of users that can be supported by a system utilizing exclusive allocation of subcarriers to users (i.e., OFDMA exclusive and FDMA). For a system employing embedded modulation with L layers and N subcarriers, the maximal number of users that can be served is $U_{max} = V = N \cdot L = 48$. This is L times more when compared to a system with exclusive allocation of subcarriers to users.

From Figure 6.25, it can be observed that for algorithms employing adaptive allocation of subcarriers to users (i.e., OFDMA embedded and OFDMA exclusive), the total transmit power requirement actually decreases with a higher number of users, assuming a fully loaded system. This is caused by the fact that for more users, the so-called *multiuser diversity* improves and the adaptive algorithms have more options to choose from in order to find a good solution.

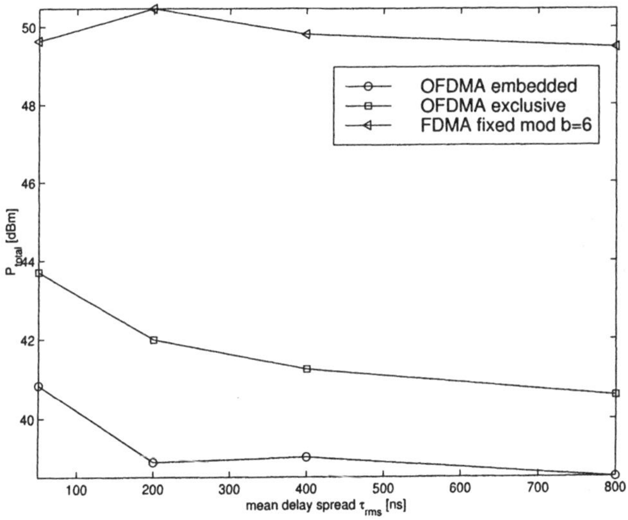

Figure 6.26 Influence of the mean rms delay spread τ_{rms}. The other system parameters are $N = 16$ for $\tau_{rms} = \{50, 200\}$ ns and $N = 32$ for $\tau_{rms} = \{400, 800\}$ ns, $U = 4$, $\alpha = 4.4$, $\sigma^2_{shadow} = 18$ dB, $Pe = 10^{-4}$. System is fully loaded.

Another interesting observation is that the performance gain of the OFDMA embedded scheme over OFDMA exclusive increases with the number of users. This is again because more users lead to higher multiuser diversity, which, in the case of OFDMA embedded, adds not only to adaptive subcarrier allocation but also to a better match of the component signals creating the embedded constellation.

6.5.7 Influence of delay spread

Comparison results for various mean rms delay spread $\tau_{rms} = \{50, 200\}$ ns for $N = 16$ and $\tau_{rms} = \{400, 800\}$ ns for $N = 32$ are depicted in Figure 6.26. It can be observed that the performance of the adaptive schemes (OFDMA embedded and OFDMA exclusive) improves with increased τ_{rms}. This is again due to higher channel variability, which allows for a better exploitation of the experienced multiuser diversity. The best performing scheme is OFDMA embedded, while the worst is FDMA.

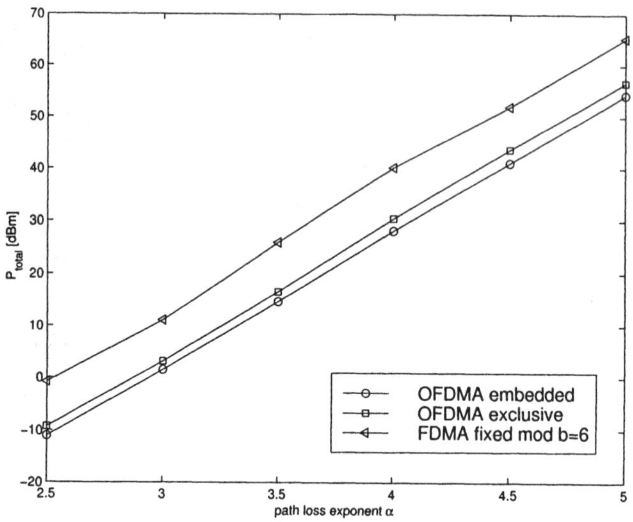

Figure 6.27 Influence of the path loss exponent α. The other system parameters are $N = 16$, $U = 4$, $\tau_{rms} = 264$ ns, σ^2_{shadow} = 18 dB, $Pe = 10^{-4}$, and the system is fully loaded.

6.5.8 Influence of path loss

Comparison results for various path loss conditions $\alpha = \{2.5, 3, 3.5, 4, 4.5, 5\}$ are depicted in Figure 6.27. The results indicate that the path loss exponent α influences all considered schemes in the same way, leading to a higher transmit power required for higher α.

6.5.9 Influence of shadow fading variance

Comparison results for different shadowing variance $\sigma^2_{shadow} = \{10, 14, 18, 22\}$ dB are shown in Figure 6.28. As can be observed, a higher shadowing variance results in a higher required transmit power.

6.6 SUMMARY

In this chapter, we have considered the problem of radio resource allocation including subcarrier assignment, bit loading, and power allocation, for high capacity

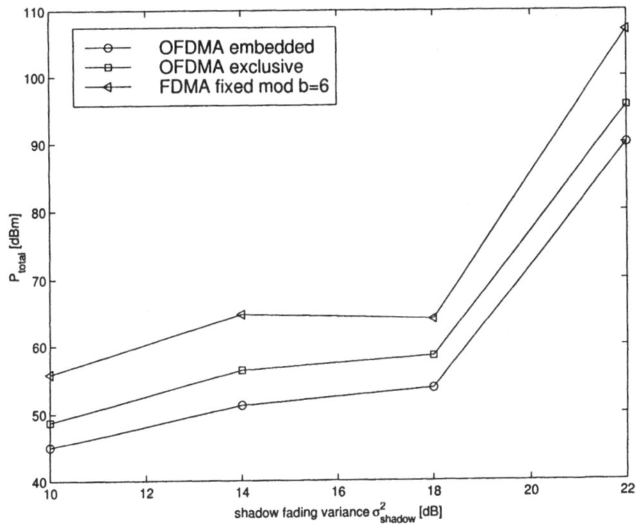

Figure 6.28 Influence of the shadow fading variance σ^2_{shadow}. The other system parameters are $N = 16$, $U = 4$, $\tau_{rms} = 264$ ns, $\alpha = 4.4$, $Pe = 10^{-4}$, and the system is fully loaded.

downlink OFDMA system with intra-cell reuse of subcarriers. After presenting a literature review setting the context for the considered system, we introduced a simplified uplink model where a subcarrier is shared between two users. These initial results indicated potential of stacking signals belonging to different users, which was followed further in this chapter when considering downlink OFDMA system model. Several algorithms were presented and evaluated, which allow for efficient distribution of the available radio resources according to the required QoS. The best performing set of algorithms is referred to as OFDMA embedded. Its performance, under various environmental conditions, has also been verified thoroughly in this chapter.

Thanks to the application of embedded modulation allowing to stack up to L, M-QAM signals ($M = 4$) on one OFDM subcarrier, the system using OFDMA embedded can offer up to L times more independent links, which may be used for serving a higher number of users. On the other hand, embedded modulation in the OFDMA context may be perceived as a generalization of adaptive modulation, where bits loaded on one subcarrier may be addressed to different users. This makes embedded modulation a more power-efficient scheme than the conventional adaptive modulation used in the OFDMA context.

We have observed that the OFDMA embedded scheme outperforms in all the considered cases of all the other reference schemes, including OFDMA exclusive (OFDMA with adaptive subcarrier allocation and adaptive modulation but with exclusive assignment of subcarriers to users, described in Chapter 5). Moreover, its performance (as well as the performance of OFDMA exclusive) improves with higher multiuser diversity, which may be caused either by more users present in the system or a higher frequency selectivity. These phenomena are disastrous for fixed schemes such as FDMA, which neglect the information about channel conditions and which apply neither adaptive subcarrier assignment nor bit loading.

References

[1] Pietrzyk, S., and Janssen, G.J.M., "Subcarrier and Power Allocation for QoS-aware OFDMA Systems Using Embedded Modulation," *Proc. IEEE International Conference on Communications*, pp. 3202-3206, June 2004, Paris, France.

[2] Cover, T.M., "Broadcast Channels," *IEEE Transactions on Information Theory*, January 1972.

[3] Ramchandran, K., Ortega, A., Metiz, Uz K., and Vetterli, M., "Multiresolution Broadcast for Digital HDTV Using Joint Source-Channel Coding," *Proc. IEEE ICC*, pp. 556-560, 1992.

[4] Ramchandran, K., Ortega, A., Metiz, Uz K., and Vetterli, M., "Multiresolution Broadcast for Digital HDTV Using Joint Source/Channel Coding," *IEEE Journal on Selected Areas in Communications*, January 1993, volume 11, pp. 6-22.

[5] Janssen, G.J.M., and Slimane, S.B., "Symbol Error Probability Analysis of a Multiuser Detector for M-PSK Signals Based on Successive Cancellation," *IEEE Journal on Selected Areas in Communications*, February 2003, volume 20, pp. 330-338.

[6] Pradhan, S. S., and Ramchandran, K., "Optimized Embedded Multicarrier Modulation for Efficient Delivery of Layered Video Data," *Proc. IEEE ICC*, pp. 1009-1012, 1998.

[7] Pradhan, S.S., and Ramchandran, K., "Efficient Layered Data Transport Over Multicarrier Systems Using Optimized Embedded Modulation," *IEEE Transactions on Communications*, June 2002, volume 50, pp. 877-881.

[8] Kim, D., and Stuber, G.L., "Performance of Multiresolution OFDM on Frequency-Selective Fading Channels," *IEEE Transactions on Vehicular Technology*, September 1999, volume 48, pp. 1740-1746.

[9] Cheng, R.S., and Verdu, S., "Gaussian Multiaccess Channels with ISI: Capacity Region and Multiuser-WaterFilling," *IEEE Transactions on Information Theory*, May 1993, volume 39, pp. 773-785.

[10] Diggavi, S. N., "Multiuser DMT: A Multiple Access Modulation Scheme," *Proc. IEEE Globecom*, pp. 1566-1570, November 1996.

[11] Zeng, C., Hoo, L.M.C., and Cioffi, J.M., "Optimal Water-filling Algorithms for a Gaussian Multi-access Channel with Intersymbol Interference," *Proc. IEEE ICC*, pp. 2421-2427, 2001.

[12] Ohno, S., Anghel, P.A., Giannakis, G.B., and Luo, Z.Q., "Multi-Carrier Multiple Access is Sum-Rate Optimal for Block Transmissions over Circulant ISI Channels," *Proc. IEEE ICC*, pp. 1656-1660, 2002.

[13] Hoo, L.M.C., Tellado, J., and Cioffi, J.M., "Multiuser Loading Algorithms for Multicarrier Systems With Embedded Constellations," *Proc. IEEE ICC*, pp. 1115-1119, 2000.

[14] Sari, H., "Orthogonal Waveform Multiple Access with Sequential Resource Sharing," *Proc. IEEE Globecom*, pp. 2380-2384, 1999.

[15] Sari, H., Vanhaverbeke F., and Moeneclaey M., "Channel Overloading in Multiuser and Single-User Communications," *Proc. IEEE PIMRC*, pp. 1106-1111, 2000.

[16] Sari, H., Vanhaverbeke, F., and Moeneclaey, M., "Multiple Access Using Two Sets of Orthogonal Signal Waveforms," *IEEE Communications Letters*, January 2000, volume 4, pp. 4-6.

[17] Sari, H., Vanhaverbeke, F., and Moeneclaey, M., "An Excess Signaling Technique with Coding, Signal Superposition, and Joint Detection," *Proc. IEEE Globecom*, pp. 926-930, 2001.

[18] Mazzini, G., "Power Division Multiple Access," *Proc. IEEE Int'l Conference on Universal Personal Communication (ICUPC)*, pp. 543-546, 1998.

[19] Janssen, G.J.M., "A Power Efficient Compound Modulation Scheme for Addressing Multiple Users in the Downlink," *Proc. 24th Symposium on Information Theory in the Benelux*, May 2003, Veldhoven, the Netherlands.

[20] Halonen, T., Romero, J., Melero, J., (eds.), *GSM, GPRS and EDGE Performance : Evolution Towards 3G/UMTS*, Wiley, 2003.

[21] Reininger, D., Izmailov, R., Rajagopalan, B., Ott, M., and Raychaudhuri, D., "Soft QoS Control in the WATMnet Broadband Wireless System," *IEEE Personal Communications*, February 1999, pp. 34-43.

[22] Janssen, G.J.M., *Robust receiver techniques for interference-limited radio channels*, Delft University of Technology, PhD Dissertation, 1998.

[23] Moretti, M., Pezzano, U.M., Sparreboom, D., and Janssen, G.J.M., "Performance Evaluation of a Mobile Communication System Implementing the Dual-Signal Receiver," *Proc. IEEE VTC'99*, Fall, pp. 482-486, 1999.

[24] Cioffi, J.M., *A Multicarrier Primer*, Amati Communications Corporation and Stanford University, November 1991.

[25] Kuhn, H.W., "The Hungarian Method for the Assignment Problem," *Naval Research Logistics Quarterly* 2, 1955.

[26] Dondeti, V.R., and Emmons, H., "Max-Min Matching Problems with Multiple Assignments," *Journal of Optimization Theory and Applications,* November 1996, volume 91, pp. 491-511.

[27] Wong, C.Y., Tsui, C.Y., Cheng, R.S., and Letaief, K.B., "A Real-time Sub-carrier Allocation Scheme for Multiple Access Downlink OFDM Transmission," *Proc. IEEE VTC,* Fall, pp. 1124-1128, 1999.

[28] Yin, H., and Liu, H., "An Efficient Multiuser Loading Algorithm for OFDM-based Broadband Wireless Systems," *Proc. IEEE Globecom,* pp. 103-107, 2000.

Chapter 7

RRM algorithms for OFDMA with inter-cell spectrum reuse

The majority of contemporary wireless access systems are cellular point-to-multipoint systems (such as GSM and UMTS [1, 2]), where the same spectrum is reused in different cells resulting in an increased system capacity and wider area coverage [3]. The same principle may be also utilized for OFDMA systems, where inter-cell reuse of subcarriers is applied. In fact, such systems are a basic prerequisite to have OFDMA applied widely as a technique for the next generation networks [4].

In this chapter, the downlink of a multicell (cellular) OFDMA system with reuse of subcarriers between different cells is considered. Recently developed RRM algorithms are described, which allow for efficient assignment of users to cells, subcarriers to users, adaptation of modulation modes (referred to as bit loading), and power control for cellular OFDMA systems. The resource allocation is performed according to users' individual QoS requirements, which are satisfied and *maintained* at the lowest possible cost of resource utilization. In addition, the performance of the described schemes is evaluated under various environmental conditions. The results reveal a high potential of these schemes in application to resource distribution in cellular OFDMA systems.

This chapter is organized as follows. In Section 7.1 we provide an overview of literature and research work on resource allocation for cellular OFDMA. Section 7.2 presents the considered system model and formulates the allocation problem. In Section 7.3, a simplified two-cell system model is considered, allowing to better understand the principles of adaptive resource allocation for cellular OFDMA systems. Section 7.4 presents a set of efficient resource allocation algorithms, while

Section 7.5 verifies their performance by simulations. Finally, conclusions are given in Section 7.6.

7.1 LITERATURE REVIEW

One of the first resource allocation methods for the cellular OFDM context is the one presented in [5] by Chuang et al. The method is based on the idea of dynamic channel allocation (DCA) [6] where each user is allowed to access any radio channel (a combination of time slot and carrier frequency) provided that the C/I of that channel can be tolerated. Thus, in principle, the entire bandwidth is available in every cell (i.e., a reuse of one is applied). However, the arrival of a new user may cause quality degradation to all cochannel users. Moreover, though the considered system utilizes OFDM, the spectrum allocation is not done on a per-subcarrier basis. Also, the QoS model is fairly simple, expressed by a variable number of traffic slots to be assigned to a user and a common target C/I. In [7], Lei and Slimane discussed a similar approach, which also utilizes DCA: a user is allocated subcarriers for which the SIR is acceptable. Neither adaptive modulation nor power control is applied. Various QoS profiles can be obtained by aggregating various number of subcarriers.

In [8], a modification of a well-known water-filling algorithm taking into account CCI is introduced by Yih and Geraniotis. Since single-cell water-filling aims at loading more power to subcarriers that are of good quality in a multicell environment with CCI, this would cause excessive interference to cochannel cells. Thus, a modification is proposed based on removing the most interfered subcarriers from the active set of subcarriers in a cell. The algorithm is executed independently at every base station. The drawback is that protection of the QoS of the cochannel users is not an explicit target of this algorithm. Moreover, TDMA is applied to separate users so in a certain time slot a user is assigned all the possible subcarriers (as determined by the water-filling algorithm). This also means than no multiuser diversity is exploited.

In [9] Zhang et al. considered adaptive subcarrier allocation in the context of CCI. The algorithm does not incorporate any notion of keeping SINR in the cochannel cells unchanged (thus protecting QoS of already established links). The proposal is very close to [10], both in using equal transmit power per each subcarrier and in the applied allocation routine. In [9] the allocation objective is to maximize the sum rate subject to individual user rate requirements, while in [10] it is to maximize minimal capacity (in order to be fair for weak users as well). Weak users in [9] (e.g., those usually located at the cell borders) have the chance to be served by a neighbor cell, since interbase station macrodiversity is considered. The algorithm chooses modulation levels according to channel conditions and the available power. Allocation of subcarriers to users is based on selecting the best subcarriers until a particular rate requirement is satisfied.

Yet another paper considering adaptive subcarrier allocation under CCI is [11], by Li et al. Also here, no means are considered to maintain and protect QoS of the cochannel users. The influence of CCI is limited by the application of a DCA-like interference avoidance scheme, where adaptive subcarrier allocation and sectorization is combined with the "virtual cell" idea, where neighboring sectors of different base stations form a "virtual cell." The allocation of resources is performed independently per each "virtual cell." The objective is to minimize the total power subject to individual data rate and error rate constraints. The proposed algorithm is also based on the two-phase routine of (a) resource allocation, and (b) allocation of subcarriers to users as in [12–14].

A mature consideration of a problem of bit loading, subcarrier assignment, and power control for OFDMA under CCI maintaining and protecting QoS of all cochannel links can be dated back to 2001. It is interesting to note that this research direction started not in the wireless domain but in the context of copper digital subscriber line (DSL) access networks. In [15] Yu et al. posed the problem of sum rate maximization, which is solved by the application of game theory and the proposed "iterative water-filling" algorithm. The CCI is handled by applying power control; no multiuser detection is exploited. In [16], Cherubini combined power control with multiuser detection. The objective of power control is to minimize the total power. The suggested solution to the posed optimization problem is claimed to be optimal and is based on simulated annealing. In [17], Sonalkar and Applegate considered the problem of sum rate maximization. The interference environment here is created by having a spectrally overlapped duplex in a DSL system. The problem is solved by transforming it to an Integer Linear Programming problem, which is then solved by a branch and bound method. In [18, 19] Lee et al. considered two closely related problems: minimization of total power [18] and maximization of the sum rate [19]. The bit loading and subcarrier allocation are based on a greedy approach in which a bit is assigned to a user and subcarrier for which it causes minimal total power increase (over all cochannel subcarriers). A solution very similar to [18] is proposed in [20] by Kulkarni and Srivastava. Also here, a greedy approach assigning a bit to the cheapest user and subcarrier taking into account all cochannel signals is used. Two algorithms are proposed, where a user utilizes (a) as many subcarriers as possible to lower sum power, and (b) as few subcarriers as possible to limit interference. The system scenario is a fixed point-to-point ad-hoc wireless network. Also, a solution similar to [18, 20] is proposed in [21] by Sonalkar. Here, in the context of overlapping duplex frequency bands of DSL with near-end cross talk (NEXT) interference, the bit and power allocation algorithm is presented. The principle is to assign a bit to a subcarrier that requires the least cost, where the cost is the sum of transmit powers at this subcarrier in both directions; upstream and downstream. Looking from a multiuser perspective, the duplex channel corresponds to only two users sharing the available spectrum. Another paper addressing a duplex channel is [22], authored by Chung and Cioffi. The objective here is to maximize sum rate, subject to individual power constraints of the two users communicating via the same set of subcarriers. The proposed algorithm is iterative in nature ("each user transmits a constant power over the selected frequency tones iteratively, presuming the interference from other users as noise") and is closely related to the one presented in [15]. Due to the two-user case, the algorithms are well applicable to cable systems (DSL).

In [23] Pietrzyk and Janssen modified a single-cell subcarrier allocation algorithm [24,25] operating under no CCI to a cellular system where CCI is present. These ideas were further continued in [26] where additionally adaptive modulation was considered in a cellular point-to-multipoint OFDMA system. The proposed methods allow for efficient radio resource utilization under the requirement to support and maintain QoS of all links.

The presented papers can be divided into three subgroups. In the first subgroup [5, 7–9, 11], though the cellular scenario is considered, CCI is not efficiently handled and QoS is not maintained, since the allocation of a user in a given cell may disrupt the QoS of all the cochannel links. In the second subgroup [15–22] CCI is handled in such a way that the QoS of all cochannel links can be maintained and protected. However, these papers do not consider a cellular scenario. All of them, except [20] have set their scenario as a DSL system. Some approaches [15, 17, 22] aim at sum rate maximization disregarding the users' individual needs. Others propose complex solutions based on optimization theory [16, 17]. The two most practical approaches seem to be [18, 20]. Both give very similar algorithms; however in both cases there is a predetermined allocation of transmitters to receivers, which is characteristic for point-to-point networks such as DSL. This does not apply to cellular networks, where a user may appear at any location within the system service area and part of the resource allocation task is to find a suitable serving base station. This is addressed in the third subgroup [23, 26], where heuristic RRM algorithms are proposed for point-to-multipoint cellular systems and CCI is handled by means of power control in such a way that it does not disrupt the QoS of any link.

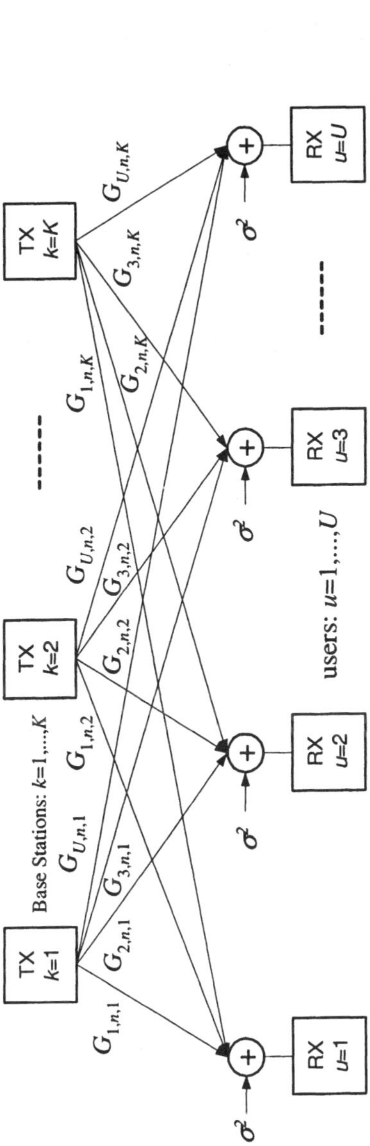

Figure 7.1 Point-to-multipoint (cellular) downlink OFDMA system with inter-cell reuse of subcarrier n, where $n = 1, \ldots, N$. Dotted lines indicate possible options for connecting BS transmitters (TX) with user terminal receivers (RX). (*After:* S. Pietrzyk and G.J.M. Janssen [26].)

In this chapter we describe RRM problems and solutions (algorithms) for cellular point-to-multipoint OFDMA systems reusing subcarriers between cells. The presented methods are able to both handle CCI and maintain QoS, and they are on purpose designed for cellular systems. This chapter is an extension of our previous work, recorded in [23, 26, 27].

7.2 SYSTEM MODEL AND PROBLEM FORMULATION

The cellular OFDMA system is composed of K base stations (BS), each operating one cell (omnidirectional antennas are used). There are the total of U users (directional antennas pointing at the BS are used) distributed uniformly over the system's service area. We consider the downlink only, where base stations communicate with users' terminals via N OFDM subcarriers, which compose the total system bandwidth of BW. It is assumed that subcarriers can be reused in different cells, which means that the same subcarrier can be allocated to different users provided that they are located in different cells. Subcarriers cannot be reused within the same cell, where they must be allocated exclusively. Also, a user cannot be served by more than one base station, either via different or the same subcarrier(s). System model for cellular OFDMA is depicted in Figure 7.1, while Figure 7.2 shows example distribution of users over system service area composed of $K = 19$ cells.

Reusing subcarriers in different cells means that, in principle, each cell can benefit from the availability of the total bandwidth BW. This is described by a so-called reuse factor of one or a so-called *single frequency network*, which is characteristic for WCDMA systems such as UMTS [1]. Reuse factor of one means that frequency planning is not necessarily what simplifies the network deployment process. In practice however, and as appears later in this chapter, the availability of particular subcarriers from the total bandwidth BW is subject to traffic, interference, and channel conditions.

It is assumed that each OFDM subcarrier is flat-fading, which is achieved by a proper selection of the subcarriers' separation Δf. Channel gains between BS k and user u communicating via subcarrier n are described by $\boldsymbol{G} = [G_{u,n,k}]$, which includes frequency selective fading, path loss, and shadowing. Due to the fact that subcarriers can be reused, cochannel interference is present. It is described by the CCI power $\boldsymbol{I} = [I_{u,n,k}]$, which indicates the total interference power received by user u served by BS K using subcarrier n. The CCI power is defined as follows

$$I_{u,n,k} = \sum_{i \neq k, \nu \neq u} G_{u,n,i} P_{\nu,n,i}, \; u = 1, \ldots, U, \tag{7.1}$$

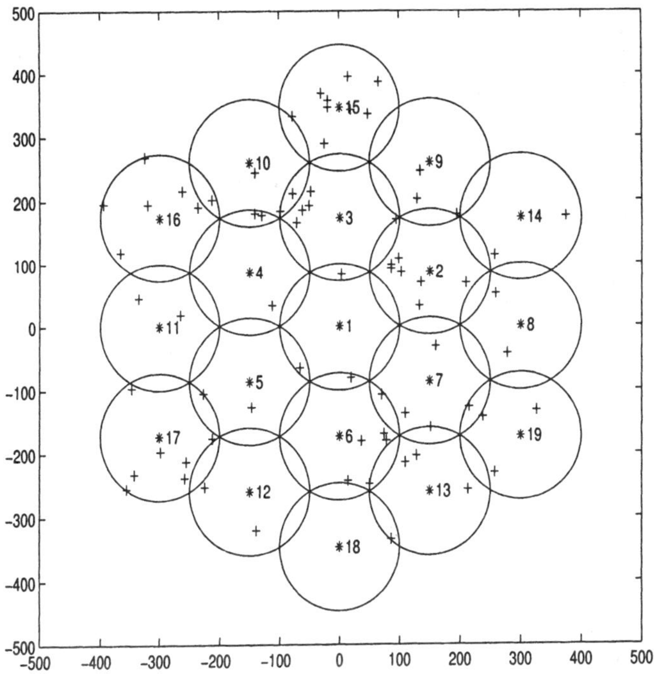

Figure 7.2 Example pattern of $K = 19$ base stations (indicated by stars), cells (indicated by circles), and $U = 95$ users (indicated by crosses).

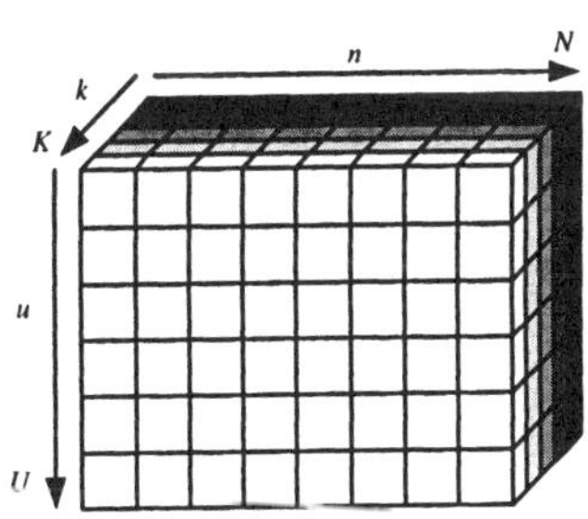

Figure 7.3 A three-dimensional representation of the following arrays: channel gain $\boldsymbol{G}$, interference $\boldsymbol{I}$, subcarrier allocation $\boldsymbol{C}$, power allocation $\boldsymbol{P}$, and bit allocation $\boldsymbol{B}$. Each dimension corresponds to $u = 1, \ldots, u$ users, $n = 1, \ldots, N$ subcarriers and $k = 1, \ldots, K$ cells, respectively.

where: $G_{u,n,i}$ indicates the channel gain between the ith interfering BS and user u and $P_{\nu,n,i}$ is the transmit power of BS i at subcarrier n that is serving user $\nu \neq u$. In addition to CCI, users' signals are corrupted by AWGN noise, which is assumed to be zero-mean with a variance σ^2. $\boldsymbol{G} = [G_{u,n,k}]$, $\boldsymbol{I} = [I_{u,n,k}]$ and σ^2 are together referred to as channel state information (CSI). It is assumed that a system (resource manager) has a perfect knowledge of CSI.

Subcarriers are allocated to users in cells according to a 3-dimensional subcarriers' allocation array $\boldsymbol{C} = [C_{u,n,k}]$, which is shown in Figure 7.3. In this array, $C_{u,n,k} = 1$ means that subcarrier n is allocated or $C_{u,n,k} = 0$ means that it is not allocated to user u served by BS k. Although it is also possible to read allocation of users to cells from $\boldsymbol{C}$, for this purpose (and for clarity) we use a separate allocation matrix $\boldsymbol{A} = [A_{u,k}]$. In this matrix, $A_{u,k} = 1$ means that a user u is allocated to a cell k, while $A_{u,k} = 0$ means that a user u is not allocated to a cell k. To express bit allocation we use bit allocation array $\boldsymbol{B} = [b_{u,n,k}]$. In this array, $b_{u,n,k}$ indicates the number of bits (per symbol) loaded on a subcarrier n allocated to user u and served by BS k. The bit loading is equivalent to setting appropriate modulation level M. In this book, for practical reasons, we restrict modulation schemes to the following M-QAMs: 4-QAM (QPSK), 16-QAM, and 64-QAM. Thus, $b_{u,n,k}$ can take values from the set $\{2, 4, 6\}$, where in addition we define the minimal possible number of bits as $b_{min} = 2$ and the maximal possible number of bits as $b_{max} = 6$. And the last resource type to be distributed is the transmit power. Its allocation is expressed by the array $\boldsymbol{P} = [P_{u,n,k}]$. In this array, given entry $P_{u,n,k}$ indicates the transmit power level that needs to be emitted from BS k using subcarrier n in order to serve user u. When, for a particular BS $\overline{k}$ and a particular subcarrier $\overline{n}$, $P_{u,\overline{n},\overline{k}} = 0$, $u = 1, \ldots, U$ then it means that the $\overline{k}$th BS is idle at subcarrier $\overline{n}$. When, $P_{u,\overline{n},\overline{k}} > 0$, $u = 1, \ldots, U$ then it means that the $\overline{k}$th BS is active at subcarrier $\overline{n}$.

As in the previous chapters, also here we assume fixed BWA system, with no (or low) user mobility. In such a scenario, the Doppler spread can be neglected. In addition, we assume that the system is perfectly synchronized (both in the time and the frequency domains).

7.2.1 Power control relations

In this subsection we explain the basics of the applied power control scheme. The power control scheme proposed in [28, 29] is applied in the considered cellular OFDMA system. A review of other power control schemes that can be found in the literature is given in the appendix at the end of the book.

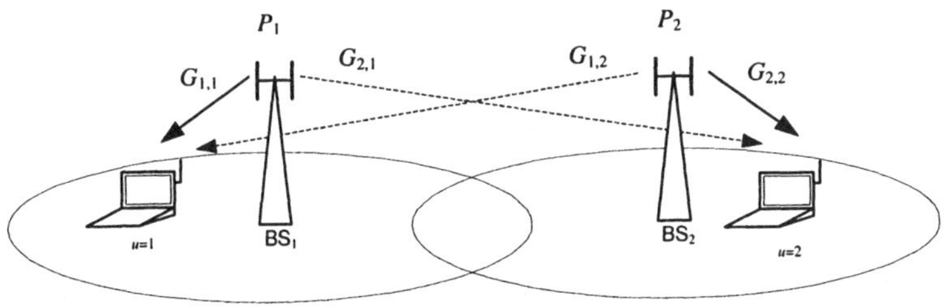

Figure 7.4 Two users reusing a common frequency channel in two different cells.

7.2.1.1 General Bambos-like power control

Let us assume a simple cellular system with two cells, two users, and one channel available for communication (e.g., a subcarrier), as depicted in Figure 7.4. It is assumed that BS $k = 1$ communicates with user $u = 1$ while BS $k = 2$ communicates with user $u = 2$ over the same frequency channel. Transmit power of BS $k = 1$ is indicated by P_1, while of $k = 2$ by P_2. Clearly, the transmission addressed for user $u = 1$ is interference to user $u = 2$ and vice versa. In order to allow both users to communicate, the following relations must hold

$$\begin{cases} SNIR_1 = \frac{P_1 G_{1,1}}{P_2 G_{1,2} + \sigma 2} \geq \gamma_1 \\ SNIR_2 = \frac{P_2 G_{2,2}}{P_1 G_{2,1} + \sigma 2} \geq \gamma_2 \end{cases}, \tag{7.2}$$

where $SNIR_1$ and $SNIR_2$ are signal-to-noise-plus-interference ratios (SNIR) experiences by user $u = 1$ and $u = 2$, respectively; γ_1 and γ_2 are minimum required SNIR thresholds, which are functions of the applied modulation level and the required bit error rate.

Rearranging (7.2) leads to the following set of linear inequalities

$$\begin{cases} P_1 G_{1,1} - \gamma_1 P_2 G_{1,2} \geq \gamma_1 \sigma^2 \\ P_2 G_{2,2} - \gamma_2 P_1 G_{2,1} \geq \gamma_2 \sigma^2 \end{cases} \tag{7.3}$$

The goal here is to find such P_1 and P_2 that satisfy the above relations, given $G_{1,1}, G_{1,2}, G_{2,1}, G_{2,2}, \gamma_1, \gamma_2$. This set can be written in a matrix form as

$$\mathbf{Hp} \geq \sigma^2 \gamma, \tag{7.4}$$

where

$$\mathbf{H} = \begin{bmatrix} G_{1,1} & -\gamma_1 G_{1,2} \\ -\gamma_2 G_{2,1} & G_{2,2} \end{bmatrix} \tag{7.5}$$

and

$$\mathbf{p} = [\ P_1 \quad P_2\ ], \tag{7.6}$$

$$\gamma = [\ \gamma_1 \quad \gamma_2\ ]^T. \tag{7.7}$$

Since we are only interested in all-positive vector of transmit powers $\mathbf{p}$, such an all-positive solution is referred to as the *feasible* one. It has been shown in [28,29] that in order to find a feasible solution to (7.2)-(7.7) it is sufficient to solve the following set of linear equations, using for example the well-known Gaussian elimination,

$$\mathbf{Hp} = \sigma^2\gamma, \tag{7.8}$$

If the solution to (7.8) is all-positive then it also satisfies (7.4). Otherwise, there is no feasible solution, which indicates that users $u = 1$ and $u = 2$ cannot share the same frequency channel because mutual interference would be too high.

7.2.1.2 Bambos-like power control in cellular OFDMA

When considering power control for cellular OFDMA, the system model presented in the previous subsection needs to be extended to account for the fact that not one but N frequency channels (in this case OFDM subcarriers) are available and the power control procedure should be applied to each of them. This also gives more freedom in resource allocation, since users may satisfy their traffic needs by having allocated multiple subcarriers. In addition, users are not preallocated to certain BS as in Subsection 7.2.1.1 but may choose the most suitable BS adaptively. Therefore, the relations presented in Subsection 7.2.1.1 need to be extended to account for a 3-dimensional nature of the allocation process, where users $u = 1, \ldots, U$, subcarriers $n = 1, \ldots, N$ and cells (BSs) $k = 1, \ldots, K$ are involved.

The generalized relation (7.2), when applied to a cellular OFDMA system takes the following form

$$SNIR_{u,n,k} = \frac{G_{u,n,k}P_{u,n,k}}{I_{u,n,k} + \sigma^2} \geq \gamma_{u,n,k}. \tag{7.9}$$

This should hold for every BS k transmitting with the use of subcarrier n.

After rearrangement of (7.9) we obtain the following set of linear inequalities, which should be fulfilled for every active cell k at subcarrier n assigned to user u

$$G_{u,n,k}P_{u,n,k} - \gamma_{u,n,k} \sum_{i \neq k, \nu \neq u} G_{u,n,i}P_{\nu,n,i} \geq \sigma^2\gamma_{u,n,k}. \tag{7.10}$$

This can also be expressed in matrix notation as follows:

$$\mathbf{H}(n)\,\mathbf{p}(n) \geq \sigma^2 \gamma(n)\,, \; n = 1, \ldots, N, \tag{7.11}$$

where

$$\mathbf{H}(n) = \begin{bmatrix} G_{1,n,1} & -\gamma_{1,n,2}G_{1,n,2} & \cdots & -\gamma_{1,n,\kappa}G_{1,n,\kappa} \\ -\gamma_{2,n,1}G_{2,n,1} & G_{2,n,2} & \ddots & \vdots \\ \vdots & \ddots & \ddots & \vdots \\ -\gamma_{\kappa,n,1}G_{\kappa,n,1} & \cdots & \cdots & G_{\kappa,n,\kappa} \end{bmatrix} \tag{7.12}$$

and

$$\mathbf{p}(n) = [p_1(n), p_2(n), \ldots, p_\kappa(n)]\,, \tag{7.13}$$

$$\gamma(n) = \begin{bmatrix} \gamma_{1,n,1} & \gamma_{2,n,2} & \cdots & \gamma_{\kappa,n,\kappa} \end{bmatrix}^T, \tag{7.14}$$

where κ is the number of BSs transmitting using subcarrier n. Also here, the goal is to find, for every subcarrier n, an all-positive vector $\mathbf{p}(n)$, given $\mathbf{H}(n)$, $\gamma(n)$, and σ^2. This vector would determine the transmit powers of all BSs using given subcarrier n. As already explained in Subsection 7.2.1.1, the solution to (7.11) can be found by solving the following equation

$$\mathbf{H}(n)\,\mathbf{p}(n) = \sigma^2 \gamma(n)\,, \; n = 1, \ldots, N. \tag{7.15}$$

This can also be solved by a well-known Gaussian elimination method. If all the elements of the transmit power vector $\mathbf{p}(n)$ are positive then this vector is referred to as feasible solution and it also solves (7.11).

Let us now define the SNIR threshold $\gamma_{u,n,k}$, as [30]

$$\gamma_{u,n,k} = \Gamma_{u,n,k}(2^{b_{u,n,k}} - 1), \tag{7.16}$$

where, assuming an uncoded system, the SNR gap $\Gamma_{u,n,k}$ is defined using the Q-function introduced in (5.6), as [30]

$$\Gamma_{u,n,k} = \frac{1}{3}\left[Q^{-1}\left(\frac{Pe_u}{4}\right)\right]^2, \; n = 1, \ldots, N; \; k = 1, \ldots, K, \tag{7.17}$$

where Pe_u is the bit error probability required by user u and $b_{u,n,k}$ is the number of bits loaded on subcarrier n as allocated to user u, which is served by cell k.

So far, we have not addressed the issue of which subcarrier should be allocated to which user and in which cell. In order to do so, first the cost array $\boldsymbol{P} = [P_{u,n,k}]$ containing the transmit power values corresponding to possible allocations $\boldsymbol{C} = [C_{u,n,k}]$ should be determined. This can be done using the following relation:

$$P_{u,n,k} = \begin{cases} p_k(n), & \text{if } k \text{ is active at } n \text{ and } u = \{u|C_{u,n,k} = 1\} \\ 0 & \text{otherwise} \end{cases} . \tag{7.18}$$

7.2.2 Resource allocation problems

In this subsection we present two resource allocation problems corresponding to the two downlink cellular OFDMA system cases, namely, constant bit rate (CBR) and variable bit rate (VBR). Analogical problems for a single-cell system without spectrum reuse have been considered in Section 5.2.2. In addition, we provide modification of these problems in order to make them applicable for the uplink scenario.

7.2.2.1 Constant Bit Rate system

The goal of the resource allocation algorithm for a CBR cellular downlink OFDMA system is to determine subcarrier allocation $\boldsymbol{C} = [C_{u,n,k}]$, bit allocation $\boldsymbol{B} = [b_{u,n,k}]$ and transmit power allocation $\boldsymbol{P}_{min} = \min(\boldsymbol{P}) = [P^{min}_{u,n,k}]$ arrays, under the constraint that the user's traffic requirements are met at the cost of minimum total transmit power. The CBR cellular allocation problem can be formulated as follows:

$$\min \sum_{u=1}^{U} \sum_{n=1}^{N} \sum_{k=1}^{K} C_{u,n,k} P_{u,n,k} \tag{7.19}$$

subject to

$$R_u \leq oR_u = \frac{1}{T_e} \sum_{n=1}^{N} \sum_{k=1}^{K} C_{u,n,k} b_{u,n,k};\ u = 1, \ldots, U, \tag{7.20}$$

$$P_{max} \geq oP_k^{total} = \sum_{u=1}^{U} \sum_{n=1}^{N} C_{u,n,k} P_{u,n,k};\ k = 1, \ldots, K, \tag{7.21}$$

$$\sum_{u=1}^{U} C_{u,n,k} \leq 1;\ n = 1, \ldots, N; k = 1, \ldots, K, \tag{7.22}$$

$$C_{u,n,k} \in \{0,1\},\ u = 1,\ldots,U;\ n = 1,\ldots,N;\ k = 1,\ldots,K, \tag{7.23}$$

where T_e is effective OFDM symbol time duration. Constraint (7.20) indicates that the offered user data rate oR_u should not be lower than required data rate R_u, and constraint (7.21) controls that the total "offered" transmit power oP_k^{total} is not higher than allowed transmit power level P_{max}. Allocation of a subcarrier to only one user within a given cell is controlled by constraint (7.22).

The optimization problem (7.19)-(7.23) for a CBR downlink cellular OFDMA system is a 3-dimensional nonlinear and combinatorial allocation problem. This problem cannot be solved as classical assignment or matching problems [31, 32] because here entries in the cost array $\boldsymbol{P} = [P_{u,n,k}]$ are mutually dependent. The cost entries $P_{u,n,k}$ for a particular cell k depend on the allocations made for all cochannel cells $i \neq k$, $i = 1,\ldots,K$ due to CCI. This is similar as in the case of resource allocation problem for single-cell OFDMA with intra-cell subcarrier reuse (6.21)-(6.25). However here, no hierarchy is applied between signals originating from different base stations. Analogically to (6.21)-(6.25), Problem (7.19)-(7.23) can be explained by the following real-life example. Let us assume that users $u = 1,\ldots,U$ correspond to tasks, subcarriers $n = 1,\ldots,N$ to workers, and cells $k = 1,\ldots,K$ to projects. $\boldsymbol{P} = [P_{u,n,k}]$ indicates the cost of worker n realizing task u within project k. Workers are specialists and they are not allowed to be assigned to different tasks within the same project. However, as often happens in real life, because there are not that many specialists available, they are allowed to be assigned to different tasks simultaneously. Clearly, the cost of realizing a given task within a given project depends both on the worker's ability to realize this task and also on his or her workload in the other projects. Therefore, we are unable to determine the cost of task u realized by worker n within project k without knowing the task's allocation in all the other projects $i \neq k, i = 1,\ldots,K$ where this worker is engaged. Occasionally it may happen that a worker is not able to be assigned a new task since he or she is already fully loaded. In such a situation acceptance of this new task would lead to quality degradation of tasks realized in the other projects, which is something that the system does not allow. The goal of the allocation problem here is to find such distribution of tasks $u = 1,\ldots,U$ among workers $n = 1,\ldots,N$ within projects $k = 1,\ldots,K$ so that the required quality of work is achieved at the lowest possible labor cost.

7.2.2.2 Variable Bit Rate system

In the VBR downlink cellular OFDMA system the goal of the resource allocation algorithm is to determine such allocation arrays $\boldsymbol{C} = [C_{u,n,k}]$, $\boldsymbol{B} = [b_{u,n,k}]$ and

$\boldsymbol{P}_{min} = \min(\boldsymbol{P}) = [P_{u,n,k}^{min}]$ so that the total data rate achieved within the entire system is maximized under the constraint that each user obtains a required minimum data rate and that available total transmit power is not exceeded at any base station. The VBR cellular allocation problem can be formulated as follows:

$$\max \sum_{u=1}^{U} \sum_{n=1}^{N} \sum_{k=1}^{K} C_{u,n,k} b_{u,n,k} \tag{7.24}$$

subject to

$$R_u \leq oR_u = \frac{1}{T_e} \sum_{n=1}^{N} \sum_{k=1}^{K} C_{u,n,k} b_{u,n,k}; \ u = 1, \ldots, U, \tag{7.25}$$

$$P_{max} \geq oP_k^{total} = \sum_{u=1}^{U} \sum_{n=1}^{N} C_{u,n,k} P_{u,n,k}; \ k = 1, \ldots, K, \tag{7.26}$$

$$\sum_{u=1}^{U} C_{u,n,k} \leq 1; \ n = 1, \ldots, N; k = 1, \ldots, K, \tag{7.27}$$

$$C_{u,n,k} \in \{0, 1\}, \ u = 1, \ldots, U; \ n = 1, \ldots, N; \ k = 1, \ldots, K. \tag{7.28}$$

The meaning of the constraints is the same as in (7.19)-(7.23). Also, a similar real-life analogy can be applied. The difference is in the allocation objective, which is here to maximize the total performance of work, given a certain price that can be paid and assuring the minimum performance level for each task.

7.2.2.3 Modification for the uplink

In order to apply problems (7.19)-(7.23) and (7.24)-(7.28) to the uplink scenario, constraints (7.21) and (7.26) have to be exchanged for the following relation:

$$P_u^{max} \geq oP_u^{total} = \sum_{n=1}^{N} C_{u,n,k} P_{u,n,k}; \ u = 1, \ldots, U; \ k = 1, \ldots, K. \tag{7.29}$$

This new constraint takes care of the individual transmit power consumption of each user.

In the following, only the CBR allocation problem (7.19)-(7.23) is further considered. It is solved by the suboptimal heuristic algorithm, which is described in Section 7.4. With minor modifications, this approach can also be applied to solve the VBR problem (7.24)-(7.28).

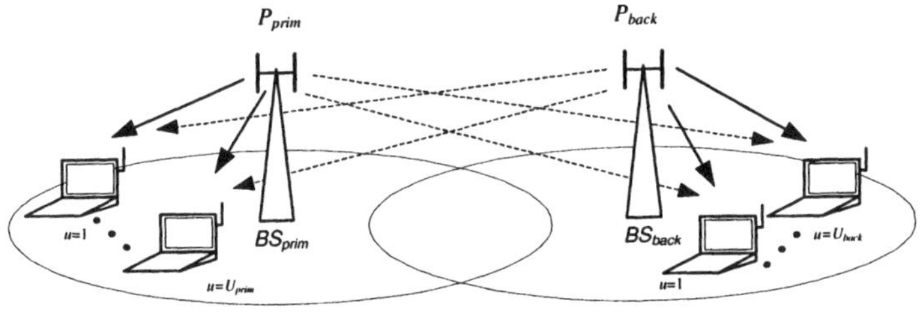

Figure 7.5 Two-cell OFDMA system scenario. Solid line indicates desired signal, dashed lines indicate interference.

7.3 SIMPLIFIED TWO-CELL MODEL

Before presenting in detail the final heuristic solution to the optimization problem (7.19)-(7.23), a simplified model consisting of only two cells is considered. This model allows to better understand the principles of adaptive resource allocation for cellular OFDMA systems with inter-cell reuse of subcarriers. In addition, it provides simulation results, which exhibit strong potential of combining adaptive subcarrier allocation with power control providing and *maintaining* the required link quality of all the cochannel users.

This section is organized as follows. Subsection 7.3.1 describes the model of the simplified two-cell system indicating the differences to the system model presented in Section 7.2 and formulates the resource allocation problem. Subsection 7.3.2 provides a solution to this allocation problem, which combines adaptive subcarrier allocation and power control efficiently handling CCI. Simulation results and final conclusions are given in Subsection 7.3.3.

7.3.1 Problem formulation

The simplified two-cell system model, as shown in Figure 7.5, is based on the model introduced in Section 7.2, except for the following:

- Only fixed modulation is allowed. This results in a constant number of bits b per subcarrier. Various data rates per user are realized by allocating various numbers of subcarriers according to user's requirements,
- Only two cells are considered: the one that triggers resource (re)allocation when experiencing channel and/or traffic changes is referred to as the *primary* cell, $k = prim$, while the other, where radio links are already established and should

be protected is referred to as the *background* cell, $k = back$; any of the two cells of the system can become primary or background,

- Allocation of users to cells given by the allocation matrix $\boldsymbol{A}$ is known apriori, the number of users in the primary cell is indicated by U_{prim} and in the background by U_{back}, total number of users is $U = U_{prim} + U_{back}$,
- No transmit power limitation is applied.

In such a system scenario, the allocation problem, (7.19)-(7.23) from Subsection 7.2.2 can be reformulated into the following form. Considering the primary cell, the goal of the resource allocation algorithm is to find such subcarrier allocation matrix $\boldsymbol{C}_{prim} = [C_{u,n,prim}]$; $u = 1, \ldots, U_{prim}$; $n = 1, \ldots, N$ for the primary cell and the corresponding power allocation matrices for the primary cell $\boldsymbol{P}_{prim}^{min}$ and the background cell $\boldsymbol{P}_{back}^{min}$ so that the total transmit power in the primary cell is minimized and users' traffic requirements in both cells are satisfied. This can be expressed formally by the following relations

$$\min \sum_{u=1}^{U_{prim}} \sum_{n=1}^{N} C_{u,n,prim} P_{u,n,prim} \tag{7.30}$$

subject to

$$R_u \leq oR_u = \frac{b}{T_e} \sum_{n=1}^{N} \sum_{k \in \{prim,back\}} C_{u,n,k}; \; u = 1, \ldots, U, \tag{7.31}$$

$$\sum_{u=1}^{U} C_{u,n,k} \leq 1; \; n = 1, \ldots, N; k \in \{prim, back\}, \tag{7.32}$$

$$C_{u,n,k} \in \{0, 1\}, \; u = 1, \ldots, U; \; n = 1, \ldots, N; \; k \in \{prim, back\}. \tag{7.33}$$

Constraint (7.31) expresses user data rate requirement R_u versus offered data rate oR_u. It can be noted that the offered data rate depends only on the number of subcarriers allocated to a user. Constraint (7.32) indicates that a subcarrier can be allocated to at most one user within a cell. Because k is no more a dimension in the optimization problem, the problem (7.30)-(7.33) is a classical matching problem [32].

7.3.2 Solution for the two-cell model

Allocation of resources for the entire cellular system at once is highly complex and therefore unpractical in reality. Moreover, satisfying the individual user traffic requests, which are unpredictable in nature, should be synchronized with the pace of all-network resource management or the all-network resource management should be triggered by the arrival of each new user. Such a highly centralized scheme would probably be extremely difficult to implement due to the delays caused by processing time needed to solve a complex optimization problem and also due to the need to communicate the results from the central resource management unit to each cell. Therefore, in order to obtain a practical and (in a way) distributed system, and based on the above reasoning, we decided to apply the resource allocation algorithm with a per-single-cell granularity. The aim is to make optimal allocations for the primary cell in presence of interference caused by the background cell. The resource allocation in the background cell during the resource allocation in the primary cell should be adapted to this new situation (new interference conditions). We aim at minimal (re)allocations in the background cell and as such we consider adaptation of the transmit power levels.

In order to solve the optimization problem (7.30)-(7.33) presented in the previous subsection, that is to find the subcarrier allocation matrix $\boldsymbol{C}_{prim}$ for the primary cell, we need to create first the cost matrix $\boldsymbol{P}_{prim} = [P_{u,n,prim}] = [P_{prim}(u,n)]$, where $u = 1, ..., U_{prim}$ and $n = 1, ..., N$. Please note that the cell index $prim$ is only a parameter here, thus the allocation and cost matrices are expressed as functions of variables (u, n). In the cost matrix $\boldsymbol{P}_{prim}$, each entry indicates the transmit power that should be spent on subcarrier n assuming that it is allocated to user u in the primary cell as shown below

$$\begin{aligned}\boldsymbol{P}_{prim} &= \begin{bmatrix} P_{1,1,prim} & P_{1,2,prim} & \cdots & P_{1,N,prim} \\ P_{2,2,prim} & P_{2,2,prim} & \ddots & \vdots \\ \vdots & \ddots & \ddots & \vdots \\ P_{U_{prim},2,prim} & \cdots & \cdots & P_{U_{prim},N,prim} \end{bmatrix} \\ &= \begin{bmatrix} P_{prim}(1,1) & P_{prim}(1,2) & \cdots & P_{prim}(1,N) \\ P_{prim}(2,2) & P_{prim}(2,2) & \ddots & \vdots \\ \vdots & \ddots & \ddots & \vdots \\ P_{prim}(U_{prim},2) & \cdots & \cdots & P_{prim}(U_{prim},2) \end{bmatrix}.\end{aligned} \tag{7.34}$$

Filling the cost matrix $\boldsymbol{P}_{prim}$ is achieved by solving (7.15) for every possible allocation of subcarrier n to user u in the primary cell. For our system, having one primary and one background cell, the $\mathbf{H}$ matrix and the γ vector are as follows:

$$\begin{aligned} \mathbf{H}(n) &= \begin{bmatrix} G_{u,n,prim} & -\gamma_{u,n,prim}G_{u,n,back} \\ -\gamma_{\nu,n,back}G_{\nu,n,prim} & G_{\nu,n,back} \end{bmatrix} \\ &= \begin{bmatrix} G_{prim}(u,n) & -\gamma_{prim}(u,n)G_{back}(u,n) \\ -\gamma_{\nu,back}(n)G_{\nu,prim}(n) & G_{\nu,back}(n) \end{bmatrix} \end{aligned} \tag{7.35}$$

and

$$\gamma(n) = \begin{bmatrix} \gamma_{u,n,prim} & \gamma_{\nu,n,back} \end{bmatrix}^T = \begin{bmatrix} \gamma_{prim}(u,n) & \gamma_{\nu,back}(n) \end{bmatrix}^T, \tag{7.36}$$

where ν indicates a user allocated a subcarrier n in the *back* cell. Using the above input data, the solution to (7.15) is the BS transmit power vector expressed as

$$\mathbf{p}(n) = [p_{prim}(u,n), p_{\nu,back}(n)]. \tag{7.37}$$

This solution should be found for each (u, n) in the primary cell. Then, $p_{prim}(u, n)$ is used to create the cost matrix $\boldsymbol{P}_{prim}$, while $p_{\nu,back}(n)$ is the corresponding transmit power required to maintain the QoS of a link between user ν and base station *back* in the background cell. If there is no solution to (7.15) for some (u, n) then this entry is excluded from the process of subcarrier allocation and indicated as banned. No solution to (7.15) means that it is not possible to load subcarrier n as allocated to user u (under given channel and traffic conditions). In such a situation, QoS of already existing links in the background cell is maintained.

Thanks to the so-called *multiuser diversity*, that is, the fact that a subcarrier as allocated to one user may have much better channel and interference conditions than as allocated to another user, the chances for finding a feasible solution are increased. Also, exhibition of all $U_{prim}\mathrm{x}N$ of (u, n) as prospective allocation combinations allows the resource allocation algorithm to pick such combinations, which lead to minimization of the transmit power. This is another advantage of multiuser diversity in a cellular OFDMA system.

Having determined the cost matrix $\boldsymbol{P}_{prim}$, the allocation problem (7.30)-(7.33) can be solved by the application of Algorithm 5.4 (*Option 1* of the *subcarriers-to-users* allocation phase in Section 5.3), where instead of searching for the best-quality subcarriers, we search for subcarriers requiring least transmit power. This scheme is further referred to as *min TX power*.

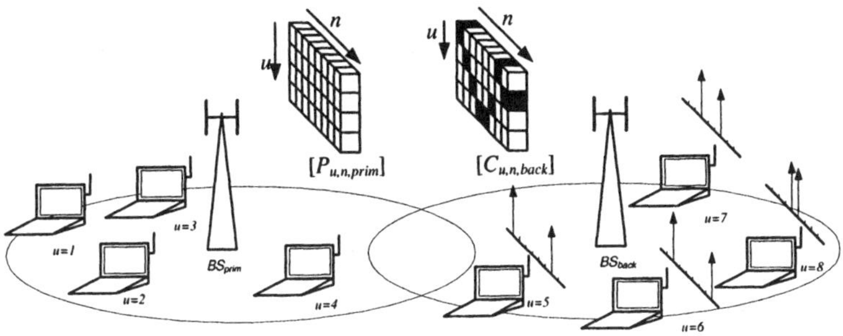

Figure 7.6 Two-cell OFDMA system with subcarriers allocated to users in the background cell according to $[C_{u,n,back}]$ and before allocation of subcarriers to users in the primary cell. Allocated entries are indicated by black elements of $[C_{u,n,back}]$.

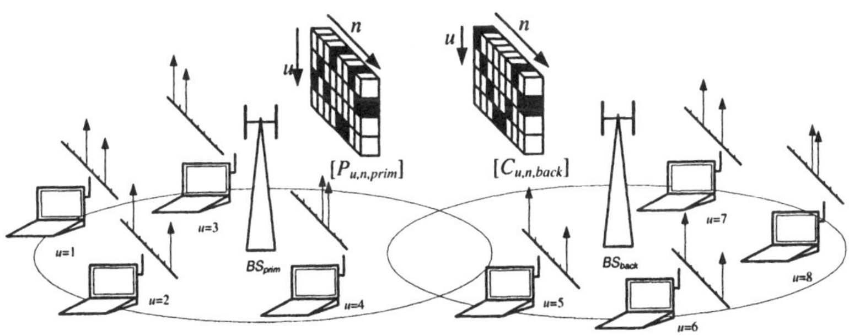

Figure 7.7 Two-cell OFDMA system with subcarriers allocated to users in the primary and background cell according to $[C_{u,n,prim}]$ and $[C_{u,n,back}]$, respectively. Allocated entries are indicated by black elements of $[C_{u,n,prim}]$ and $[C_{u,n,back}]$. Transmit power in the background cell is corrected when compared to Figure 7.6.

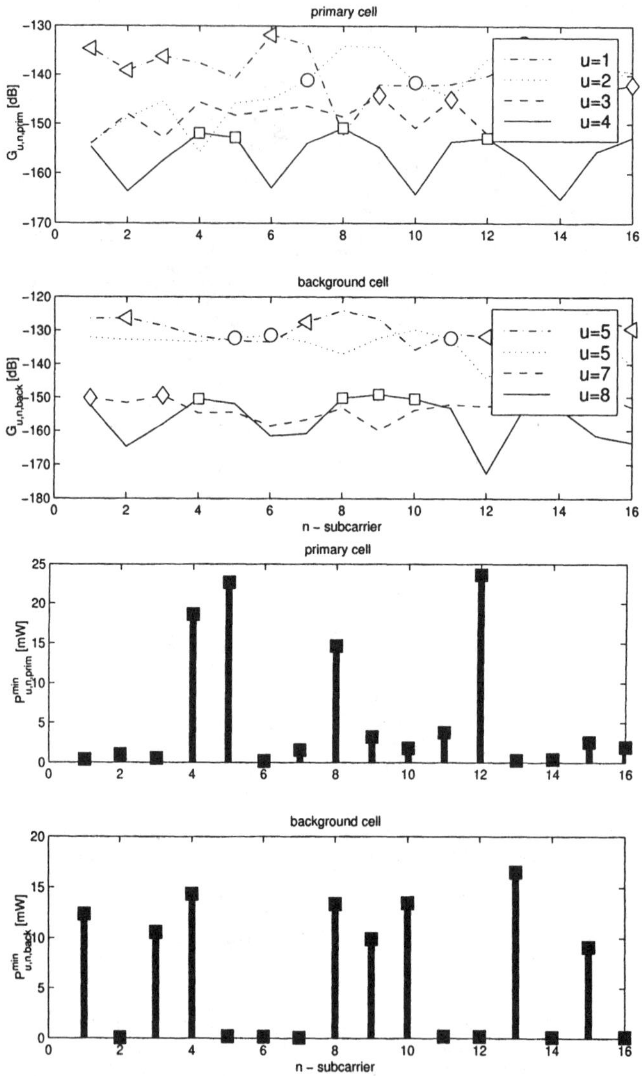

Figure 7.8 Example channel gains and subcarrier allocations (upper picture) and power allocation (lower picture) for $U = 8$, $N = 16$, $K = 2$. Each user requires a data rate corresponding to four subcarriers.

An example case of two-cell OFDMA system with $N = 8$ subcarriers and $U = 8$ users is depicted in Figures 7.6 and 7.7. In Figure 7.6 a situation before allocation in the primary cell is depicted. There, allocation of subcarriers to users in the background cell is indicated by $[C_{u,n,back}]$ matrix (allocated entries are black), which is used to indicate which interference elements should be taken into account in composing the primary cell cost matrix $[P_{u,n,prim}]$. Having $[P_{u,n,prim}]$ we can distribute subcarriers to users within the primary cell, which is depicted in Figure 7.7. The transmit power for users in the primary cell is set together with correcting the transmit power of users in the background cell in order to adapt to a new interference situation. Please note that subcarriers are reused between primary and background cell, while they are allocated exclusively to one user within one cell. Example subcarrier to user allocation and transmit power assignment for $N = 16$ is depicted in Figure 7.8. For better understanding, the upper picture of Figure 7.8 shows also channel gains corresponding to each user ($G_{u,n,prim}$ and $G_{u,n,back}$). It can be observed that the algorithm assigns subcarrier to users adaptively and tends to pick good quality subcarriers (as perceived by a given user) for allocation. Also, it can be seen that all subcarriers are reused in the two cells and CCI is controlled thanks to appropriate transmit power settings ($P^{min}_{u,n,prim}$ and $P^{min}_{u,n,back}$).

In the above approach, first (7.15) is solved for every possible (u, n) allocation in order to create the cost matrix $\boldsymbol{P}_{prim}$ and then adaptive subcarrier allocation is performed in the primary cell. This requires U_{prim}xN runs of the Gaussian elimination algorithm to find a solution to (7.15). Aiming to reduce the computational complexity, a modification of the original approach is proposed, which requires only N runs of (7.15). This reduction is achieved by first performing adaptive subcarrier allocation and then solving (7.15) only for such (u, n) allocations, which select best channel conditions in the primary cell. In order to do subcarrier allocation, the cost function $\boldsymbol{P}_{prim} = [P_{u,n,prim}]$ is replaced with the benefit function $\boldsymbol{T}_{prim} = [T_{u,n,prim}]$ determined according to

$$T_{u,n,prim} = \frac{G_{u,n,prim}}{(I_{u,n,prim} + \sigma^2)\,\Gamma_{u,n,prim}}. \tag{7.38}$$

The new allocation problem is then formulated in the following way

$$\max \sum_{u=1}^{U_{prim}} \sum_{n=1}^{N} C_{u,n,prim} T_{u,n,prim} \tag{7.39}$$

subject to

$$R_u \leq oR_u = \frac{b}{T_e}\sum_{n=1}^{N}\sum_{k\in\{prim,back\}} C_{u,n,k};\ u = 1,\ldots,U, \tag{7.40}$$

$$\sum_{u=1}^{U} C_{u,n,k} \leq 1;\ n = 1,\ldots,N; k \in \{prim, back\}, \tag{7.41}$$

$$C_{u,n,k} \in \{0,1\},\ u = 1,\ldots,U;\ n = 1,\ldots,N;\ k \in \{prim, back\}. \tag{7.42}$$

The solution to this problem is a subcarrier allocation matrix $\boldsymbol{C}_{prim}$. This can be directly obtained by the application of Algorithm 5.4. Now, for the entries indicated by ones in $\boldsymbol{C}_{prim}$, (7.15) is solved. If there is no solution to (7.15) then the original one is changed to zero in the final allocation matrix $\boldsymbol{C}_{prim}$. Otherwise, the feasible power vector (7.37) is calculated and the transmit power values are written to $\boldsymbol{P}_{prim}^{min}$ and $\boldsymbol{P}_{back}^{min}$ according to (7.18). This scheme is further referred to as *max gain*.

7.3.3 Simulation results

The performance of the described RRM algorithms for the simplified two-cell system model has been evaluated by simulations. The two algorithms introduced in Subsection 7.3.2 are referred to as *min TX power* (original) and *max gain* (modified). For comparison, the following performance metrics are used: (a) *average transmit power "per-subcarrier"* $P_{prim}^{av} = \sum_u \sum_n P_{u,n,prim}^{min} / \sum_u \sum_n C_{u,n,prim}$ for the primary cell and similarly for the background cell P_{back}^{av} and (b) *fulfillment coefficient* $Q_{prim}^{ful} = \sum_u \sum_n C_{u,n,prim}/N$ for the primary cell and similarly for the background cell Q_{back}^{ful}, assuming the requirement to utilize N subcarriers in each cell. A maximum value of 1 means that all the required subcarriers are allocated.

The performance of the described schemes (min TX power and max gain) is compared to the performance of the following reference RRM techniques:

- *Optimal*: similar to the *prop* method with the exception that the subcarriers in both cells are adaptively allocated using optimal Hungarian algorithm.
- *no PC no ad sub* (no power control, no adaptive subcarrier allocation): where power control protecting the QoS of existing links is not used (fixed transmit power level is applied for all subcarriers according to worst channel conditions) and subcarriers are assigned in a fixed manner (classical FDMA approach) to users in both cells.

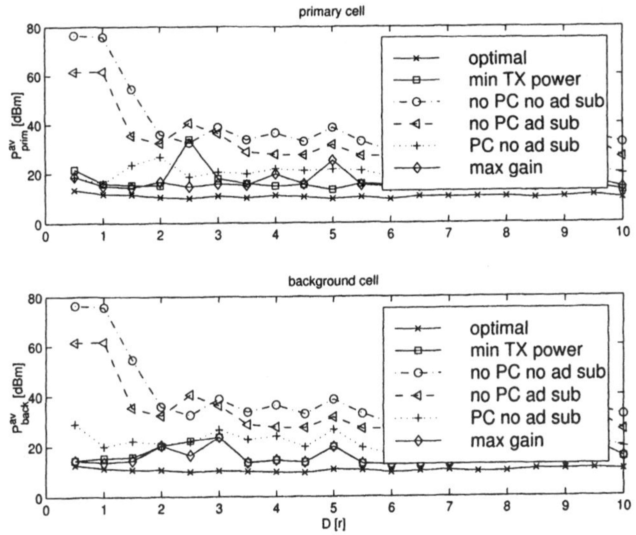

Figure 7.9 Comparison of the average transmit power "per-subcarrier" as a function of reuse distance D between centers of the two cells normalized to the cell radius r for the described algorithm (min TX power), its modified version (max gain), and reference schemes (no PC no ad sub, no PC ad sub, PC no ad sub).

- *no PC ad sub* (no power control, adaptive subcarrier allocation): where still power control is not used (also worst-case design approach is applied) but subcarriers are allocated adaptively to users.
- *PC no ad sub* (power control, no adaptive subcarrier allocation): where power control providing and maintaining QoS of all links is applied according to (7.15) but where subcarriers are assigned to users in a fixed way in both cells (classical FDMA).

The performance comparison between the presented and reference RRM schemes is shown in Figure 7.9 and Figure 7.10. Both performance metrics are depicted as functions of reuse distance D between the centers of the two cells normalized to the cell radius r. The values of the performance metrics are obtained by averaging results from 100 different channel / traffic realizations. In both cells there is the same number of users $U_{prim} = U_{back} = 8$ and each user requires a data rate equivalent to 2 subcarriers, which means that the goal is to fully load a system, since the total number of subcarriers is $N = 16$. Channel parameters are as

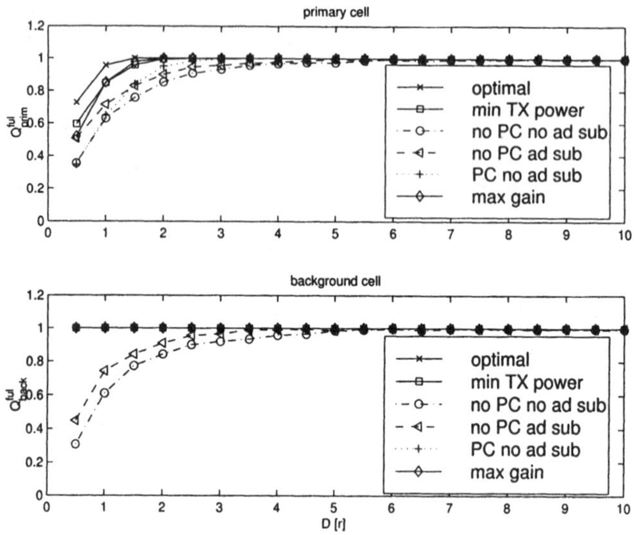

Figure 7.10 Comparison of the fulfillment coefficient as a function of reuse distance D between centers of the two cells normalized to the cell radius r.

described in Chapter 4, including mean rms delay spread $\tau_{rms} = 264$ ns, path loss exponent $\alpha = 4.4$, mean Rician $K_{factor} = -3$ dB , attenuation $PL(d_o) = -39$ dB at the reference distance of $d_o = 1$ m, and shadowing variance $\sigma^2_{shadow} = 18$ dB.

We can see from Figure 7.9 and Figure 7.10 that both of the described methods min TX power and max gain are better than the reference schemes including no PC no ad sub, no PC ad sub and lastly PC no ad sub, producing the lowest average transmit power "per-subcarrier" and highest fulfillment coefficient. In addition, the performance of both algorithms (min TX power and max gain) is very close to the *optimal* case.

As can be seen in Figure 7.10, a full reuse of subcarriers in both cells can be achieved at the distance of $2r$ for the min TX power scheme with max gain method slightly behind. This is a good result because the PC no ad sub scheme achieves the same reuse at the distance of $4r$, while schemes without power control, namely, no PC no ad sub and no PC ad sub require approximately $6r$. These results also show that it is more important, from the perspectives of *fulfillment coefficient*, to apply power control providing and protecting required QoS than to apply adaptive subcarrier allocation. This is also visible in Figure 7.9, where the scheme PC no ad sub requires less transmit power than no PC ad sub.

In both Figure 7.9 and Figure 7.10 the performance of max gain is slightly worse than the performance of original min TX power. This is of course due to the fact of limited multiuser diversity leading to less choices in finding feasible solutions. However, the max gain scheme is easier to implement and requires less computational complexity. This is because the algorithm does not have to search for a feasible solution over every possible combination of users and subcarriers but does it only for chosen pairs picked according to best channel conditions. For these reasons, the *max gain* routine forms the basis for constructing the subcarriers-to-users allocation scheme for the multiple-cell system, which is presented in the next section.

It can be observed that all schemes implementing power control (i.e., min TX power, max gain and reference PC no ad sub) do not affect the QoS of the already established links in the background cell. All the other methods degrade the QoS of already established links in the background cell until a sufficiently large distance between cells is possible. It can also be seen that the transmit power levels in both the primary and the background cells are similar. Although not shown in this book, but observed in the simulations, these levels remained stable over thousands of simulation runs. This indicates that adaptive subcarrier allocation in the primary cell does not cause sudden transmit power increase in the background cell and vice versa.

The major deliverable of this subsection treating a simplified two-cell OFDMA model, is in using the combination of (a) adaptive subcarrier allocation for minimization of the transmit power, and (b) power control to provide and protect the required QoS of all the active links. This provides a basis for multicell solutions as presented in the next section.

7.4 SOLUTION FOR A MULTICELL SYSTEM

7.4.1 Design assumptions

After considering the simplified two-cell system, we return to our original optimization problem for a multicell system (i.e., (7.19)-(7.23)), which was stated in Section 7.2. The goal of this problem is to minimize the sum transmit power of the entire system. This is a general but fair goal since it aims at the overall performance of the entire system. However, an algorithm attempting to find an optimal solution for such a goal may be difficult to implement. This is because a cellular system usually consists of a number of independent cells, each with a number of users characterized by their channel and traffic conditions. These users may access or exit the system and in addition their channel and traffic conditions may change over time. Such a scenario is clearly more dynamic than the one of fixed DSL-like systems [18, 20].

An interesting question to state is "what and how often should trigger an algorithm solving problem (7.19)-(7.23)?" Clearly, we would like to have our system adaptive to changing conditions of the environment. It is plausible to assume that in a cellular system all changing conditions are related to (or caused by) the users. After the first user is allowed to the system, we always have to deal with the environmental changes related to users already present in the system and new user(s) accessing the system. This was already indicated in the case of the simplified two-cell model described in Section 7.3 by having the primary cell, which is the one with new channel/traffic conditions and the background cell, which is the one serving users already allowed to the system. A channel/traffic change may happen to one, two, or more users at the same time instant during which we would like to distribute the resources. Because a change of channel/traffic conditions related to one user is the smallest level of granularity in the environmental changes, in the construction of RRM algorithm for cellular OFDMA further in this chapter, we take the conditions of one user as a trigger for our RRM algorithm.

Another interesting question to pose is "should all the system resources be involved in solving problem (7.19)-(7.23) after triggering by a traffic/channel change related to one user (either existing or new)?" A resource allocation system that would have to adapt to each channel/traffic change by reallocating all the network resources would be highly complex from the implementation perspective, and therefore unpractical in reality. At the other extreme, we can envision a system where (re)allocation involves only resources corresponding to this one new user accessing a system (which is often the case in contemporary systems such as GSM). Here, however, allocation of some resources for a new user may cause link degradation for the users already active in the system due to cochannel interference. For example, allocation of particular subcarriers for a new user and setting certain transmit powers causes a new interference situation in the network that influences all cochannel users and possibly degrades the performance of their radio links. This calls for some level of adaptation of allocations for both new users and users already active in the system. Moreover, when processing only one user regardless of the other users in the candidate cell, it is not possible to fully benefit from the multiuser diversity. This leads to suboptimal assignments, where a new user may be left without any option for having good subcarriers in case they are already distributed over existing users. This also calls for cell-level reallocations of resources applied to users already active in the system. The need for a cell-level allocation to exploit multiuser diversity was proven to be successful in [33].

Taking all this into account, in designing the allocation algorithm presented further in this chapter, the following directions are taken that locate the proposed solution in between the two mentioned extreme cases for resource allocation, namely (a) (re)allocation of all the network resources and (b) (re)allocation involving only a new user:

- The (re)allocation of certain system resources (but not of the entire system) is initiated (triggered) by the change of channel/traffic conditions of a single user. This change of channel/traffic conditions can be caused by the arrival of a new user and/or departure of an existing user. Nevertheless, most of the observations and conclusions presented further in this chapter hold also for a system where at one allocation run more than one user can request to access or depart a system.
- A user triggering (re)allocation of system resources has its resources allocated together with all the other users belonging to the candidate cell in order to exploit locally (i.e., within the cell) multiuser diversity.
- Similarly to the simplified two-cell model described in Section 7.3, resources in all cells other than a candidate (i.e., all the other cells that can be referred to as background) are not reallocated except for the transmit power. The transmit power is corrected in order to protect the quality of the cochannel links (i.e., those already existing in the system and the ones newly accessing). It is also important to note that a key element of admission control mechanism is checking if such setting of transmit powers of all cochannel users can be found that satisfy all users' quality demands at the acceptable price (e.g., the price can be the transmit power). If we can find such a setting then a new user is allowed to the system. Otherwise, the new user is rejected because allowing this user to the system would degrade the QoS of the other cochannel users and one of our primary objectives is to maintain the QoS of users already active in the system.

Thanks to the application of the above-presented design directions, the system is *cooperative* in the sense of resource distribution both within a cell (subcarrier allocation) and between cells (spectrum reuse). This is because users belonging to the same cell are allocated subcarriers collectively (though exclusively), while users belonging to different cells do not treat the other (cochannel) users as harmful noise, but a power control algorithm finds transmit power vectors satisfying all cochannel users. Thus, interference environment referred to as a *competitive* one in [34], can be here considered as *cooperative*.

In the following, we describe a suboptimal RRM algorithm for a cellular OFDMA system reusing subcarriers between different cells. The described solution follows the above-presented design directions and overcomes the main difficulty of the original allocation problem (7.19)-(7.23), namely the mutual dependency (due to CCI) of the entries in the cost matrix.

7.4.2 Master algorithm

For simplicity, let us consider one new user triggering the resource allocation. In a similar way to the method described in Section 5.3.2, also here the complete RRM algorithm (also referred to as the "master" algorithm) for cellular OFDMA system consists of the three steps allocating different portions of resources. The complete "master" algorithm for resource allocation triggered by the arrival of one new user is depicted in Algorithm 7.1. The three steps of this algorithm are as follows:

- *Step 1*: "user-to-base-station allocation" subalgorithm finds a candidate cell for our new user. This is done by determining the list of preferred cells, where a cell first on the list is the best candidate for a considered user, while the last one is the worst candidate. Also in this step the algorithm checks the so-called *necessary condition* for admission of a considered user to the first cell on the list. This necessary condition is simply a verification if the total data rate demand (related to our considered user and all the users already existing in this first cell) does not exceed maximal traffic that can be offered with N subcarriers. If the necessary condition is not fulfilled (i.e., total data rate demand is above capacity of all subcarriers), a next cell on the list of preferred cells is tried until we check all the cells. If it is not possible to fulfill the necessary condition in any of the cells from the list then a user is blocked. This approach is referred to as adaptive cell selection (ACS). ACS was considered in the OFDMA scenario in [9] but its forms are known in contemporary systems (e.g., cell selection/reselection routines in GSM). Thanks to ACS, if a candidate cell is unable to serve a user, another cell is tried from the list of preferred cells. In this way the chances for user admission are increased.
- *Step 2*: "subcarrier to user allocation" subalgorithm allocates subcarriers to users within a candidate cell for which the necessary condition has been fulfilled in *Step 1*. Here, subcarriers are allocated for our new user together with reallocating

Algorithm 7.1: "Master" routine for adding a new user in a cellular OFDMA system with inter-cell reuse of subcarriers.

input : R_u, resource allocations for existing users: $\boldsymbol{A}, \boldsymbol{C}, \boldsymbol{B}, \boldsymbol{P}_{min}$
output: updated resource allocations for new and existing users: $\boldsymbol{A}, \boldsymbol{C}, \boldsymbol{B}, \boldsymbol{P}_{min}$

begin
  $\boldsymbol{A}_{temp} = \boldsymbol{A}, \boldsymbol{C}_{temp} = \boldsymbol{C}, \boldsymbol{B}_{temp} = \boldsymbol{B}, \boldsymbol{P}_{min}^{temp} = \boldsymbol{P}_{min}$
  $s_u^{min} = \lceil T_e R_u / b_{max} \rceil, s_k^{min} = \sum_{u \in \text{cell } k} s_u^{min}$
  for *each new user* $\overline{u}$ *with* $R_{\overline{u}}, Pe_{\overline{u}}, G_{\overline{u},n,k}, I_{\overline{u},n,k}$ **do**
    $satisf = 0, K_{checked} = 0$
    Step 1: find a list of preferred cells $[F_i]$ for user $\overline{u}$
    $j = 1$
    while $satisf = 0$ *and* $K_{checked} < K_{checked}^{max}$ **do**
      $\overline{k} = F_j$
      if $\left(s_{\overline{u}}^{min} + s_{\overline{k}}^{min} \leq N\right)$ **then**
        update $\boldsymbol{A}_{temp}$ by setting $A_{\overline{u},\overline{k}}^{temp} = 1$
        Step 2: allocate subcarriers to users in cell $\overline{k}$, update $\boldsymbol{C}_{temp}$
        Step 3: allocate: bits in cell $\overline{k}$ and power levels in all cells; update $\boldsymbol{B}_{temp}, \boldsymbol{P}_{min}^{temp}$, and $\boldsymbol{C}_{temp}$
        if $oR_u \geq R_u$ *and* $oPe_u \leq Pe_u$ *and* $P_k^{total} \leq P_{max}, u = 1, \ldots, U$ **then**
          $satisf = 1$
        else
          $satisf = 0; K_{checked} = K_{checked} + 1; j = j + 1$
      else
        $satisf = 0; K_{checked} = K_{checked} + 1; j = j + 1$
      if $satisf = 1$ **then**
        allow user $\overline{u}$ to the system and apply:
        $\boldsymbol{A} = \boldsymbol{A}_{temp}, \boldsymbol{C} = \boldsymbol{C}_{temp}, \boldsymbol{B} = \boldsymbol{B}_{temp}, \boldsymbol{P}_{min} = \boldsymbol{P}_{min}^{temp}$
      else
        block user $\overline{u}$
end

subcarriers of all the users already active in this cell. This is done in order to exploit multiuser diversity within a cell.

- *Step 3*: "per-subcarrier bit loading and transmit power control" subalgorithm determines, for each user in the candidate cell and in all cochannel cells, the transmit power and modulation levels per each allocated subcarrier. Finding the transmit power for each subcarrier of each cochannel user is done by solving the feasibility equation (7.11)-(7.15).

After all the above-presented resource allocation steps, the algorithm verifies if the QoS objectives of all the users can be fulfilled. This constitutes *sufficient condition* for admission. If the sufficient condition is fulfilled then a new user is allowed to the system and resource distribution is executed. Otherwise, the "master" algorithm gives a chance for another cell on the list of preferred cells and the entire three-step routine is repeated. If the sufficient condition cannot be met for any of the cells on the list, a user is blocked and no resource reallocation is applied to the other users already active in the system.

Algorithm 7.2: "Master" routine for removing existing user in a cellular OFDMA system with inter-cell reuse of subcarriers.

---

input : resource allocations for existing users: $\boldsymbol{A}, \boldsymbol{C}, \boldsymbol{B}, \boldsymbol{P}_{min}$
output: updated resource allocations after removing a user: $\boldsymbol{A}, \boldsymbol{C}, \boldsymbol{B}, \boldsymbol{P}_{min}$

begin
  $\boldsymbol{A}_{temp} = \boldsymbol{A}, \boldsymbol{C}_{temp} = \boldsymbol{C}, \boldsymbol{B}_{temp} = \boldsymbol{B}, \boldsymbol{P}_{min}^{temp} = \boldsymbol{P}_{min}$
  for *each user* $\overline{u}$ *to be removed with* $R_{\overline{u}}, Pe_{\overline{u}}, G_{\overline{u},n,k}, I_{\overline{u},n,k}$ **do**
    Step 1: find cell $\overline{k}$ serving user $\overline{u}$, remove $\overline{u}$ from $\boldsymbol{A}_{temp}, \boldsymbol{C}_{temp}, \boldsymbol{B}_{temp}, \boldsymbol{P}_{min}^{temp}, \boldsymbol{G}, \boldsymbol{I}$
    Step 2: allocate subcarriers to remaining users in cell $\overline{k}$, update $\boldsymbol{C}_{temp}$
    Step 3: allocate: bits in cell $\overline{k}$ and power levels in all cells; update $\boldsymbol{B}_{temp}, \boldsymbol{P}_{min}^{temp}$, and $\boldsymbol{C}_{temp}$
    $\boldsymbol{A} = \boldsymbol{A}_{temp}, \boldsymbol{C} = \boldsymbol{C}_{temp}, \boldsymbol{B} = \boldsymbol{B}_{temp}, \boldsymbol{P}_{min} = \boldsymbol{P}_{min}^{temp}$
end

When a certain user is to be removed, Algorithm 7.2 can be applied. However, for simplicity of exposition, in this book we only consider the process of adding users.

7.4.3 Step 1: User-to-base-station allocation

The goal of Step 1 of RRM for cellular OFDMA is to suggest the best cell (base station) that should serve a new user. This is achieved by producing a list $[F_j]$ of preferred cells, where cells on the list are sorted in a descending order according to the average channel quality as perceived by user u. The list of preferred cells can be expressed as

$$F_j = \arg\ \text{sort}_k \overline{T}_{u,k}, \tag{7.43}$$

where $\overline{T}_{u,k}$ is the average (over all subcarriers) normalized channel gain for user u at cell k and is defined as

$$\overline{T}_{u,k} = \frac{1}{N}\sum_{n=1}^{N} T_{u,n,k}, \tag{7.44}$$

where $T_{u,n,k}$ is the normalized channel gain defined as

$$T_{u,n,k} = \frac{G_{u,n,k}}{(I_{u,n,k} + \sigma^2)\,\Gamma_{u,n,k}}. \tag{7.45}$$

As can be seen in (7.45), the normalized channel gain reflects not only channel gains but also interference levels and therefore is suitable for cellular scenarios with CCI.

7.4.4 Step 2: Subcarrier-to-user allocation

Having found the best cell $\overline{k}$ for user $\overline{u}$ we can now pass to Step 2, in which subcarriers are allocated to user $\overline{u}$. As already indicated, the allocation of subcarriers to user $\overline{u}$ is done together with reallocation of subcarriers to all the other users already present in cell $\overline{k}$. Such per-cell subcarrier allocation allows for exploiting multiuser diversity, which leads to more efficient resource distribution. The output of Step 2 routine is the subcarrier allocation array $[C_{u,n,k}^{temp}]$, which is indicated by *temp* to reflect that this is a temporal array, executed later (after Step 3) when the sufficient condition for admission (i.e., QoS requirements of all users) is fulfilled. Step 2 routine consists of the following two phases:

- *Phase A*, which is presented as Algorithm 7.3 and which is a slightly modified version of Algorithm 5.1 is used. In this approach the minimal number of subcarriers is initially assigned to each user in order to fulfill its rate requirement and then the remaining subcarriers are distributed over users requiring least transmit power.

Algorithm 7.3: Determination of the number of subcarriers per user for a cellular OFDMA system. Phase A of Step 2.

input : $[\overline{T}_{u,k}]$, R_u, $\overline{k}$
output: $s_u, u \in$ cell $\overline{k}$

begin
  $s_u^{max} = \lceil T_e R_u / b_{min} \rceil; s_u = s_u^{min} = \lceil T_e R_u / b_{max} \rceil; u \in$ cell $\overline{k}$
  while $\sum_{u=1}^{U} s_u < N$ *and* $s_u \leq s_u^{max}$ **do**
    for $u \in$ *cell* $\overline{k}$ **do**
      Calculate ΔP_u with (7.47)
      $w = \arg\max_u \Delta P_u$
      $s_w = s_w + 1$
      $\overline{P}_w = \overline{P}_w + \Delta P_w$
end

Algorithm 7.4: Allocation of subcarriers to users in a cellular OFDMA system. Phase B of Step 2.

input : $[T_{u,n,k}]$, s_u, $\overline{k}$
output: $[C_{u,n,k}^{temp}]$

begin
  $C_{u,n,\overline{k}}^{temp} = 0; u \in$ cell $\overline{k}, n = 1, \ldots, N$
  while $\sum_{u \in cell\, \overline{k}} s_u > 0$ **do**
    for $u \in$ *cell* $\overline{k}$ **do**
      $m = \arg\max_{n \in \{\text{not assigned}\}} T_{u,n,\overline{k}}$
      $C_{u,m,\overline{k}}^{temp} = 1$
      $s_u = s_u - 1$
end

- *Phase B*, which is presented as Algorithm 7.4 and which is a derivative of Algorithm 5.4 based on a simple greedy routine distributing subcarriers among users based on the best normalized channel gains.

The two phases are discussed in more detail below.

When compared to the original version presented as Algorithm 5.1, the algorithm for phase A is modified in order to avoid allocating more subcarriers than a user requires. Algorithm 5.1 distributes all available subcarriers among users. Such an approach, though justified for a one-cell scenario, in case of multiple cells may lead to unnecessary blocking of some subcarriers for users in cochannel cells, where those subcarriers might be better exploited. The modification is realized by applying a condition that $s_u \leq s_u^{max}$ for all users $u \in$ cell $\overline{k}$, where s_u^{max} is the maximal number of subcarriers that a user would require to satisfy its data rate demand R_u, and is defined as

$$s_u^{max} = \left\lceil \frac{T_e R_u}{b_{min}} \right\rceil, \tag{7.46}$$

where b_{min} is the minimal number of bits allowed by a modulation scheme.

As in its original version, Algorithm 7.3 assigns one new subcarrier to this user, for which it leads to maximal power reduction. This power reduction is defined as follows

$$\Delta P_u = \frac{s_u \left(2^{T_e R_u / s_u} - 1\right)}{\overline{T}_{u,\overline{k}}} - \frac{(s_u + 1)\left(2^{T_e R_u/(s_u+1)} - 1\right)}{\overline{T}_{u,\overline{k}}}. \tag{7.47}$$

After having determined the number of subcarriers s_u that each user in cell $\overline{k}$ requires the algorithm passes to phase B, which is based on Algorithm 5.4. In phase B, subcarriers are allocated to users in cell $\overline{k}$ according to the normalized channel gain matrix $[T_{u,n,\overline{k}}]$ defined in (7.45). Such a strategy is motivated by the fact that users experiencing better channel conditions would require less transmit power and would cause less cochannel interference. It proved to be a successful approach in the case of a simplified two-cell model discussed in Section 7.3. The problem of subcarrier-to-user allocation has been considered in many papers, including [13, 24, 25]. It can be solved optimally by applying the Hungarian algorithm [31]. However, here for a cellular OFDMA system, a simple greedy and suboptimal routine is presented in Algorithm 7.4, which is based on Algorithm 5.4. If necessary, the subcarrier-to-user allocation can be further improved by the use of iterative swapping of subcarriers between users, as shown in Algorithm 5.5.

After application of Algorithm 7.4, the allocation arrays $\boldsymbol{C}_{temp}$ and $\boldsymbol{A}_{temp}$ are known. They are indicated as *temp* (temporal), since the decision whether to apply them or not is still to be taken after bit loading and power control and verification of the sufficient condition for admission.

7.4.5 Step 3: Per subcarrier bit loading and transmit power control

Now, when users are temporarily allocated to cells (what is indicated by $\boldsymbol{A}_{temp}$), subcarriers to users (indicated by $\boldsymbol{C}_{temp}$), the algorithm can perform the last allocation task for a given user u that is to choose the modulation levels and adjust the transmit powers for each subcarrier allocated to user u. This is done together with adjusting the transmit powers of all cochannel users already present in the system in order to protect QoS of their links. This last step of RRM allocation for cellular OFDMA systems described in this subsection is in fact a single-user bit loading (and transmit power control) performed in the presence of background CCI. This routine is based on the ones proposed in [18,20] for point-to-point interference-limited links, which are in fact modifications of Algorithm 5.7 adapted for the operation under CCI.

The Step 3 greedy subroutine operates in the following way. For each user u assigned to cell $\overline{k}$ the algorithm searches for the cheapest (in terms of transmit power) subcarrier m chosen from the set of subcarriers allocated to a particular user u, for increase of modulation level by one order. Allowing only QPSK, 16-QAM, and 64-QAM this corresponds to: (1) starting to transmit with QPSK, (2) increasing modulation level from QPSK to 16-QAM, or (3) increasing modulation level from 16-QAM to 64-QAM. Increasing modulation level is equivalent to increasing the number of bits from $b_{u,n,l}$ to $b^{new}_{u,n,l}$. When such lowest-cost subcarrier is found, which is indicated by m, the number of bits is increased to $b^{new}_{u,m,\overline{k}}$ and the appropriate transmit power is applied to this subcarrier according to $P^{new}_{u,m,\overline{k}}$ and to all cochannel cells, also reusing subcarrier m to protect their QoS according to (7.15) and (7.18). If it happens that it is not possible to further increase the number of bits loaded on a subcarrier because, for example, the maximal possible number of bits b_{max} is already reached or there is no feasible transmit power vector $\mathbf{p}(n)$ satisfying QoS of all cochannel links on subcarrier n, then a verified triple $(u, n, \overline{k})$ is marked as *not feasible* and is not considered as a candidate for increasing a bit load. If all triples $(u, n, \overline{k})$ (i.e., all subcarriers n allocated to user u in cell $\overline{k}$) are not feasible the system is referred to as *saturated*, which causes the algorithm to exit the bit loading loop and indicates that either QoS objectives of a new user cannot be met or QoS objectives of existing users would have to be violated. In effect this leads to blocking this new user from admission to the system. However, if the system is not saturated, the per-subcarrier bit loading and transmit power control loop repeats until data rate requirements of all users belonging to cell $\overline{k}$ are satisfied without disrupting QoS of users allocated cochannel links.

It may happen for a particular user that bit loading will not utilize all the subcarriers indicated in $\boldsymbol{C}_{temp}$. In such a situation, the excess subcarriers are released and $\boldsymbol{C}_{temp}$ is updated. The resulting bit loading and power allocation, as well as updated subcarrier allocation, are part of the all-system allocation arrays $\boldsymbol{B}_{temp}$, $\boldsymbol{P}_{min}^{temp}$, $\boldsymbol{C}_{temp}$, where only the involved entries are modified. The complete Step 3 routine is shown as Algorithm 7.5.

7.5 NUMERICAL EXPERIMENTS

In the previous section we introduced algorithms for RRM in the downlink cellular OFDMA system with inter-cell reuse of subcarriers. The "master" algorithm consists of the three resource assignment steps: Step 1, Step 2, and Step 3. In the following, the name *OFDMA cellular* will refer to the complete "master" algorithm.

In this section we present results of numerical experiments evaluating the performance of the described OFDMA cellular scheme. The simulated cellular system consisted of $K = 19$ cells each with one base station as shown in Figure 7.2. Each base station is equipped with an omnidirectional antenna. Users are randomly distributed over the system area. Each user is equipped with a terminal, each having a directional antenna pointing at the base station. We use here the channel model presented in Chapter 2 and the traffic model presented in Chapter 5. Unless otherwise stated, the main system parameters are as introduced in Chapter 4, including system bandwidth BW = 7 MHz, mean rms delay spread $\tau_{rms} = 264$ ns, path loss exponent $\alpha = 4.4$, mean Rician $K_{factor} = -3$ dB , attenuation $PL(d_o) = -39$ dB at the reference distance of $d_o = 1$m, and shadowing variance $\sigma^2_{shadow} = 18$ dB. For simplicity, each user is assumed to demand the same bit rate $R_u = R$ and the same bit error rate $Pe_u = Pe$. In the simulations, although the number of cells is relatively low $K = 19$, no wraparound technique was used in order to reflect a small urban BWA hot spot.

In the following, the performance of the OFDMA cellular algorithm and its versions is compared to the performance of the other RRM schemes. This is followed by verification of the influence of changing the following system and/or environmental parameters:

- Mean rms delay spread $\tau_{rms} = \{50, 200, 400, 800\}$ ns;

Algorithm 7.5: Bit loading and power control for a cellular OFDMA system.

input : $[P_{u,n,k}]$, R_u, $\overline{k}$
output: $\mathbf{B}_{temp}$, $\mathbf{P}_{min}^{temp}$

begin
  for $u \in$ *cell* $\overline{k}$ **do**
    saturated = 0
    while $oR_u < R_u$ *and saturated* $= 0$ **do**
      for $n \in$ *user* u **do**
        if $b_{u,n,\overline{k}} = b_{max}$ **then**
          mark entry $(u, n, \overline{k})$ as not feasible
        else
          increase number of bits from $b_{u,n,\overline{k}}$ to $b_{u,n,\overline{k}}^{new}$
          calculate $P_{u,n,\overline{k}}^{current} = \sum \mathbf{p}(n, b_{u,n,\overline{k}}) = \sum_{k=1}^{K} p_k(n, b_{u,n,\overline{k}})$ and $P_{u,n,\overline{k}}^{new} = \sum \mathbf{p}(n, b_{u,n,\overline{k}}^{new}) = \sum_{k=1}^{K} p_k(n, b_{u,n,\overline{k}}^{new})$ using (7.16)-(7.15)
          if *no feasible power vector* $\mathbf{p}(n)$ *is found i.e.,* $P_{u,n,\overline{k}}^{new} = \oslash$ **then**
            mark entry $(u, n, \overline{k})$ as not feasible
      if *all considered* $(u, n, \overline{k})$ *are not feasible* **then**
        saturated = 1
      else
        $\Delta P_{u,n,\overline{k}} = P_{u,n,\overline{k}}^{new} - P_{u,n,\overline{k}}^{current}$
        $m = \arg\min_{n \in \text{user } u} \Delta P_{u,n,\overline{k}}$
        increase number of bits from $b_{u,m,\overline{k}}$ to $b_{u,m,\overline{k}}^{new}$
        update power values in cell $\overline{k}$ according to $P_{u,m,\overline{k}}^{new}$ and in all CCI cells according to $\mathbf{p}(m)$ as in (7.15)
        update $b_{u,m,\overline{k}}^{new}$ in $\mathbf{B}_{temp}$ and $P_{u,m,\overline{k}}^{new}$ with $\mathbf{p}(m)$ in $\mathbf{P}_{min}^{temp}$
end

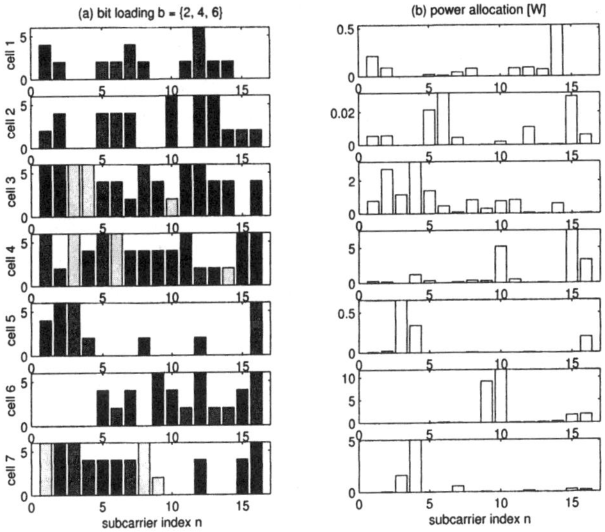

Figure 7.11 Example allocation patterns for: (a) bits and subcarriers, and (b) transmit power. The applied RRM algorithm is the OFDMA cellular. $K = 7$ cells (for clarity of exposition), $R_u = R = 6$ Mbit/s, $Pe_u = Pe = 10^{-4}$. A given user is assigned exclusively to one cell. In a given cell, different users are indicated by different gray levels.

- Data rate required by a single user $R = \{1, 3, 6, 12\}$ Mbit/s;
- Path loss exponent $\alpha = \{2, 3, 4, 5\}$;
- Variance of the Gaussian distributed power variations due to shadowing $\sigma^2_{shadow} = \{10, 14, 18, 22\}$ dB.

7.5.1 Example allocation patterns

Example bit loading and subcarrier allocation patterns are shown in Figure 7.11(a). It can be seen that for a given cell, subcarriers are allocated to users adaptively. In addition, bits are loaded on subcarriers based on the underlying channel and CCI conditions, as well as users' data rate demands. The data rate demands, which in this example case are set according to $R_u = R = 6$ Mbit/s are satisfied for all users, because each user gets at least 14 bits and symbol duration is $T_e = 2.2857$ ms, which gives $oR_u = 6.125$ Mbit/s. A very important observation is that subcarriers are reused between different cells and therefore they are mutually interfering. However this interference is tolerable, thanks to the feasible transmit power settings depicted in Figure 7.11(b) satisfying SNIR demands of all cochannel links. Again, it should be noted that a new user is not allowed to the system if it could violate the QoS requirements of existing users or the QoS of this new user could not be fulfilled.

7.5.2 Performance comparison

We have chosen the following RRM schemes for cellular OFDMA for performance comparisons:

- *Lower bound (optimal)* scheme, which is basically the exhaustive search routine performed over all possible assignments limited to the case where $K = 2$, $N = 4$, and $U = 4$ due to the excessive computational time; it is only used in Figure 7.14 and in Table 7.1.
- *OFDMA cellular*, where the following modulation levels are allowed: QPSK, 16-QAM, and 64-QAM; also adaptive cell selection (ACS) is applied and the maximal number of cells where a new user tries to be allowed to the system is $K_{checked}^{max} = 5$.
- *OFDMA cellular no ACS*, where also QPSK, 16-QAM, and 64-QAM are allowed; however ACS is not applied and thus a new user may try to be allowed to only one cell, $K_{checked}^{max} = 1$.
- *OFDMA cellular no ACS fixed mod*, where only one modulation level is used, namely QPSK, ACS is not applied, $K_{checked}^{max} = 1$.
- *FDMA*: in this reference scheme subcarriers are allocated to users according to fixed FDMA way where a new user gets a set of first unallocated subcarriers, regardless their quality. The number of subcarriers is determined based on the user's data rate requirement, which also determines the modulation level according to 2^b-QAM, with fixed number of bits b. SNIR of cochannel links is provided and maintained by the use of power control, which is the same as in case of OFDMA cellular. ACS is not applied, $K_{checked}^{max} = 1$.

The following metrics have been used as means to compare performance of the considered RRM algorithms and other reference schemes:

- Total data rate offered, defined as the sum of all users' offered data rate $oR_{total} = \sum_u oR_u$;
- Blocking probability P_{block} defined as the probability that a new user is blocked;

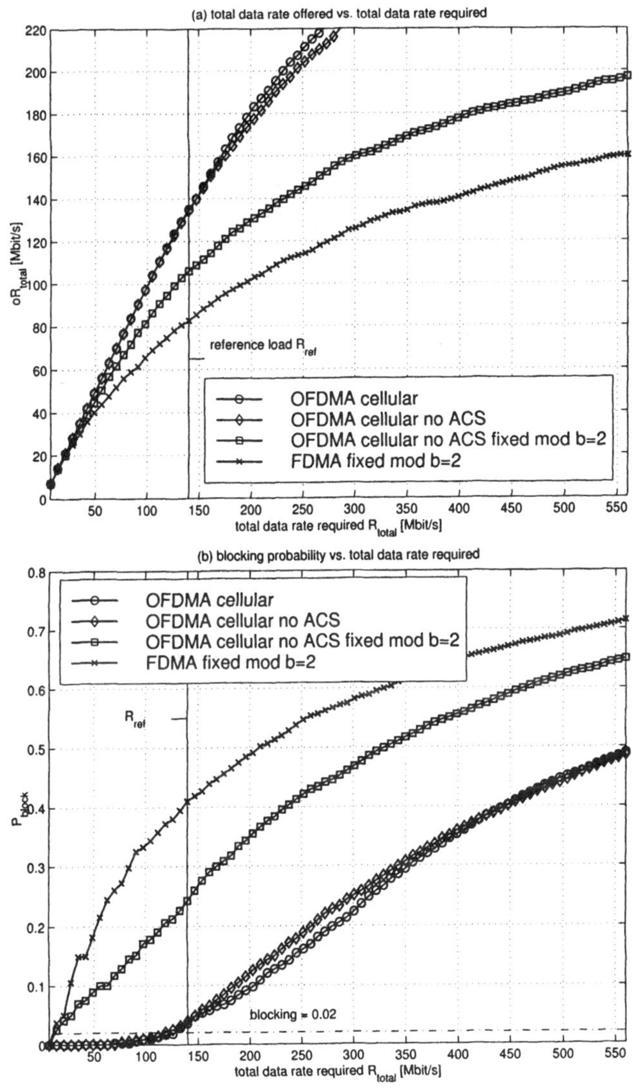

Figure 7.12 Performance comparison curves for different RRM algorithms applied in a downlink cellular OFDMA system (parts (c) and (d) are presented on the next page). $R_u = R = 6$ Mbit/s, $Pe_u = Pe = 10^{-4}$.

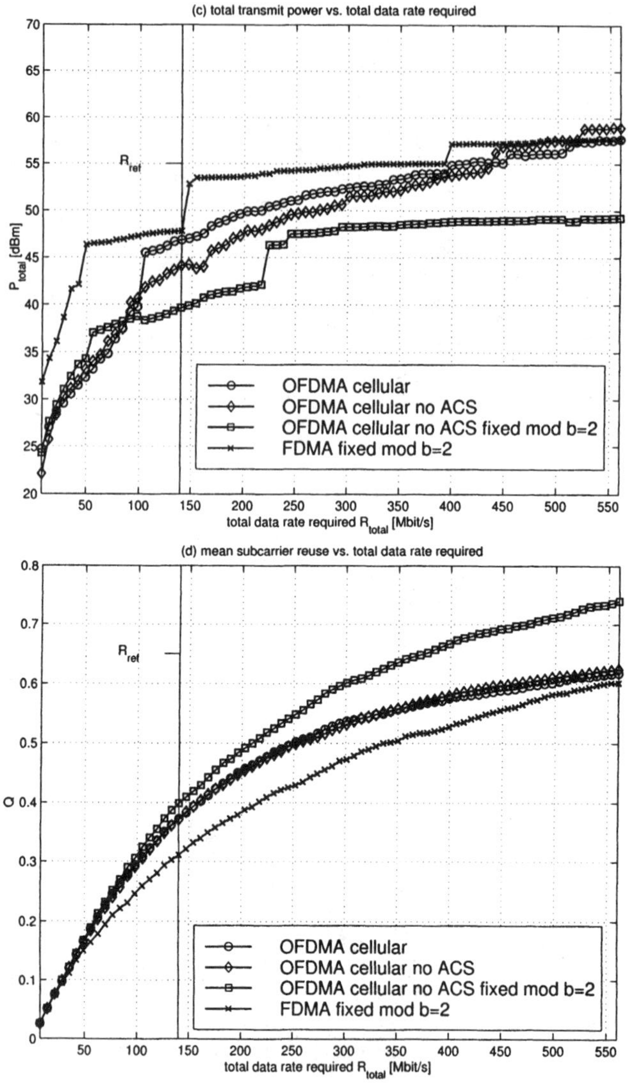

Figure 7.12 (cont.)

- Total system transmit power, which is given by $P_{total} = \sum_u \sum_n \sum_k P^{min}_{u,n,k}$;
- Mean subcarrier reuse Q, which is defined as $Q = \sum_n \kappa_n / KN$, where κ_n is the number of cells where subcarrier n is allocated. This measure shows how aggressively a subcarrier is reused: a mean subcarrier reuse equal to 1 indicates that, on average, every subcarrier is reused in every cell.

The results of the performance comparisons are depicted in Figure 7.12. In all of the four parts, (a)-(d), the performance metrics are presented as a function of total data rate required R_{total}. This reflects the way in which the system is filled with traffic: users are generated one-by-one in random locations of the system service area, as explained in the simulation flow chart shown in Figure 7.13. Then, resource allocation algorithm is executed (e.g., the "master" routine, Algorithm 7.1), which finds new resource distribution in order to allow a new user to the system. If available resources are insufficient, a new user is blocked, otherwise it is admitted to the system. Performance comparison metrics are then averaged over Z (e.g., $Z = 100$) different channel traffic realizations to make the results statistically valid. The total data rate required R_{total} is a sum of required data rate of all users that ever accessed the system.

For all the considered schemes compared in Figure 7.12, we can observe that the more traffic is required, the more traffic is offered, the blocking probability is increased, the total transmit power is higher, and also the subcarriers' reuse is increased. In Figure 7.12(a) it can be seen that the offered total data rate tends to saturate, which indicates that more CCI is present in the system and the system is unable to serve all the new users. This can also be observed in Figure 7.12(b), where, for OFDMA cellular and OFDMA cellular no ACS, blocking probability increases rapidly for total data rates required higher than a certain threshold (corresponding approximately to 0.02 blocking probability). The more traffic is served, the more sum transmit power is needed, as can be seen in Figure 7.12(c). In addition, with higher traffic demands, the subcarriers' reuse improves, since more users are served with the same spectrum, which is depicted in Figure 7.12(d).

At the reference blocking probability of 0.02 (which is often used in GSM system design), the total data rate offered for the OFDMA cellular and similarly for the OFDMA cellular no ACS scheme is ~125 Mbit/s, while for the OFDMA cellular no ACS fixed mod and FDMA it is barely ~10 Mbit/s. The OFDMA cellular family is superior to the FDMA scheme thanks to the application of adaptive (per-cell) allocation of subcarriers to users, which allows to exploit multiuser diversity and, in the case of OFDMA cellular and OFDMA cellular no ACS, also thanks to the application of adaptive modulation. The advantage of OFDMA cellular over

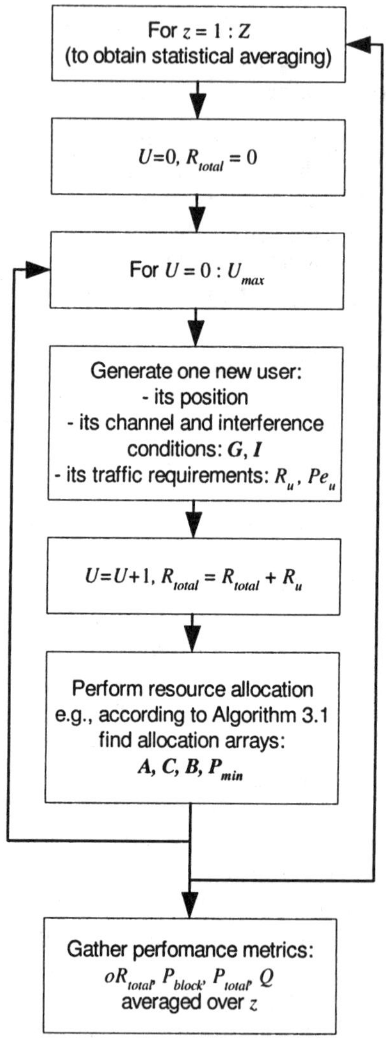

Figure 7.13 Flow chart for simulations.

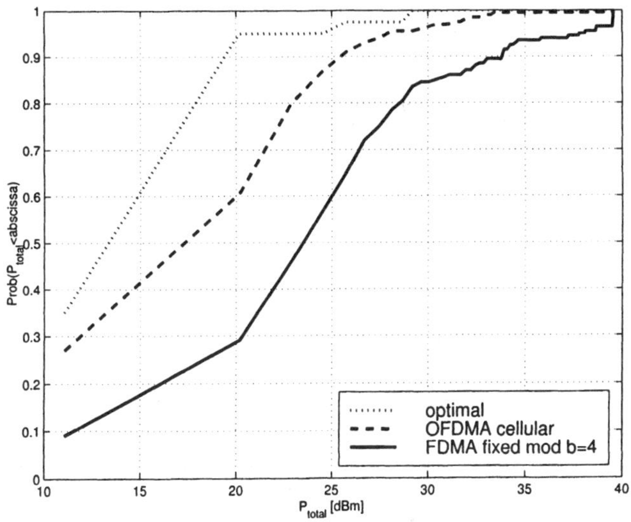

Figure 7.14 Comparison of the cumulative distribution function of the total transmit power for a cellular downlink OFDMA system. The system parameters are as follows: $R_u = R = 5$ Mbit/s, $Pe_u = Pe = 10^{-4}$, $K = 2$, $N = 4$, $U = 4$, $BW = 5$ MHz.

OFDMA cellular no ACS, visible for higher total loads in Figure 7.12(a), is due to applying ACS, which increases the chance to find a suitable cell for a user. The advantage of OFDMA cellular and OFDMA cellular no ACS over OFDMA cellular no ACS fixed mod is due to applying adaptive modulation, which allows to better exploit channel / interference variability. The price to be paid by the OFDMA cellular is relatively high total transmit power, which stems mainly from the use of adaptive modulation and application of ACS, which can be seen in Figure 7.12(c). Figure 7.12(d) indicates that the highest subcarrier reuse is observed for the OFDMA cellular no ACS fixed mod scheme. This is due to the fact that though the algorithm allocates subcarriers adaptively to users, a user may satisfy its data rate requirement only by having allocated more subcarriers and not by increasing the modulation order because the modulation order is fixed in this case. The lowest subcarriers' reuse is found for the FDMA scheme.

Performance comparison of the described OFDMA cellular scheme to the lower bound (optimal) solution and FDMA method is shown in Figure 7.14. Simulation scenario is limited here to $K = 2$, $N = 4$ and $U = 4$ due to the excessive computational time for the lower bound (optimal) scheme. Clearly,

Table 7.1

Performance Comparison of RRM Schemes Used in Figure 7.14

	optimal	*OFDMA cellular*	*FDMA* $b = 4$
P_{total} [dBm]	46.9	54.9	75.2
$oR_{total}[Mbit/s]$	20	19.4	16.9

OFDMA cellular outperforms FDMA ($b = 4$) in terms of the total transmit power P_{total}, but it is worse than the optimal case. This is summarized in Table 7.1 showing average (over different channel and traffic realizations) values of P_{total} and in addition total data rate offered oR_{total}. It is expected, however, that for higher U, N, and K the performance of OFDMA cellular will go closer to optimal. This is because exposition of more options for allocation to the adaptive OFDMA cellular algorithm results in its improved performance.

In the following, verification of the algorithms' performance under various environmental conditions is presented by making a snapshot of the performance comparison metrics (averaged over 100 different channel and traffic realizations) at a reference load of $R_{ref} = 140$ Mbit/s.

7.5.3 Influence of delay spread

In this subsection, we compare the performance of the considered schemes for various mean delay spread values, namely $\tau_{rms} = \{50, 200, 400, 800\}$ ns. The results of performance evaluation are depicted in Figure 7.15 (a)-(d). The best performance in terms of the highest offered data rate and the lowest blocking probability is observed for the OFDMA cellular scheme. As already indicated, the price for its superiority is higher total transmit power, Figure 7.15(c). However, it is interesting to note that the total transmit power for OFDMA cellular is decaying in the fastest way, among all the OFDMA-based methods, with the increased values of τ_{rms}. Also, the blocking probability for all the OFDMA-based schemes (OFDMA cellular, OFDMA cellular no ACS and OFDMA cellular no ACS fixed mod) improves (lowers) with larger τ_{rms} (thus increased frequency selectivity). This is because OFDMA allows for adaptive allocation of subcarriers to users thus exploiting multiuser diversity. The more frequency selectivity is present in the system, the more chances we have to get diversified subcarriers' qualities as perceived by various users and thus chances for more efficient allocations

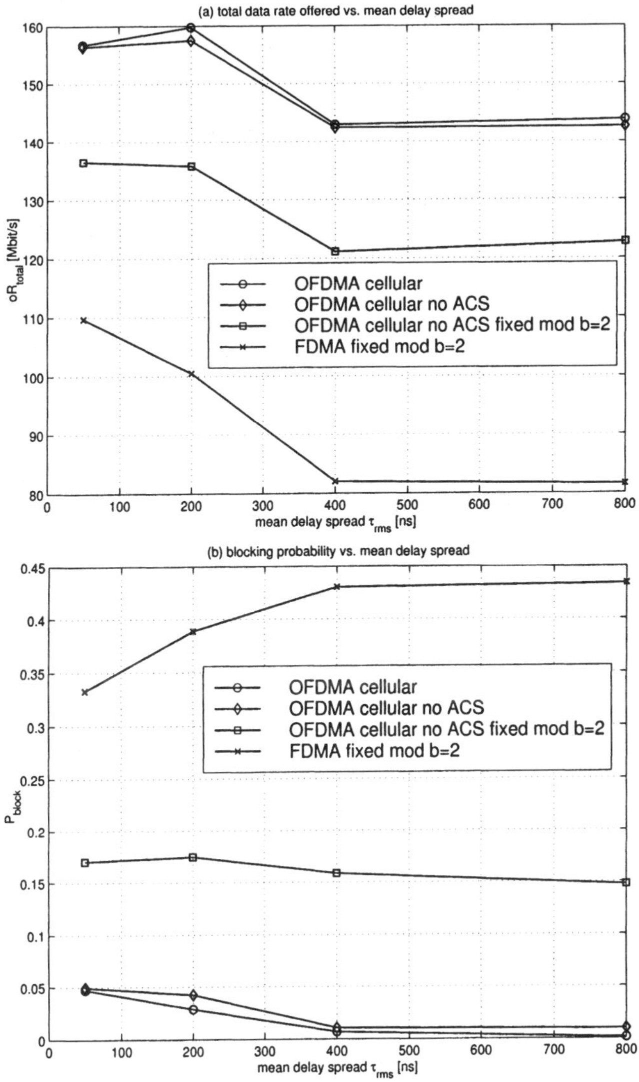

Figure 7.15 Influence of the mean delay spread τ_{rms}. The other system parameters are $N = 16$ for $\tau_{rms} = \{50, 200\}$ ns and $N = 32$ for $\tau_{rms} = \{400, 800\}$ ns, $\alpha = 4.4$, $R = 3$ Mbit/s, σ^2_{shadow} = 18 dB.

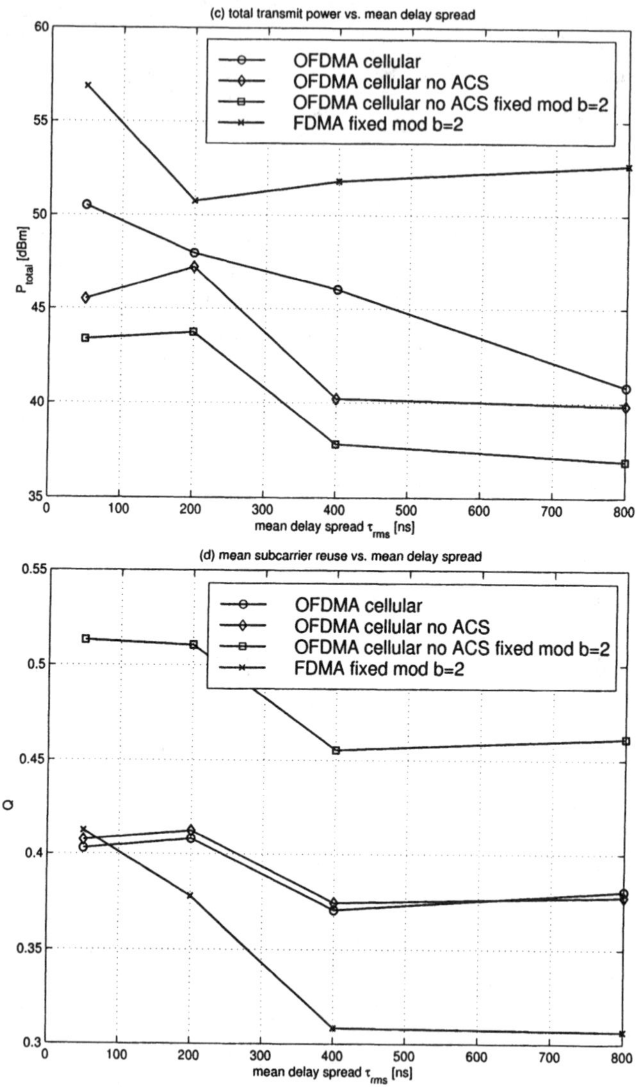

Figure 7.15 (cont.)

increase. Only for FDMA, which does not benefit from the multiuser diversity, the blocking probability is increased with higher τ_{rms}. Regarding the subcarriers' reuse, OFDMA cellular is, together with OFDMA cellular no ACS in the middle between FDMA with its lowest reuse and OFDMA cellular no ACS fixed mod with its highest subcarriers' reuse.

The total data rate offered, as shown in Figure 7.15(a) for all the verified schemes, degrades with increased τ_{rms}. This is due to the fact that the granularity of the offered data rates is determined by the number of assigned subcarriers and chosen modulation levels (from three available levels). For $\tau_{rms} = \{50, 200\}$, the number of subcarriers is lower (i.e., $N = 16$) and thus the granularity is lower than for $\tau_{rms} = \{400, 800\}$, where $N = 32$. For $N = 16$, users are offered higher data rates than they really require (i.e., more than 3 Mbit/s per single user) because offering exactly the required 3 Mbit/s is not possible. In the case of $N = 32$, granularity is higher and it is possible to offer data rates closer to the demanded 3 Mbit/s. It is important to note that the total data rate offered for OFDMA cellular and OFDMA cellular no ACS is above the required reference load of 140 Mbit/s. This does not hold for a reference OFDMA cellular no ACS fixed mod and FDMA, which are unable to offer such traffic load.

7.5.4 Influence of data rate required by a single user

In this section we validate the influence of various data rates required by a single user, namely $R_u = R = \{1, 3, 6, 12\}$ Mbit/s. The simulation results are shown in Figure 7.16 (a)-(d). Also here, similarly as in the previous section, the total data rate offered drops with increased R but is still above the threshold 140 Mbit/s for OFDMA cellular and OFDMA cellular no ACS until $R_u = 6$ Mbit/s. This degradation is again due to discrete values of offered data rates as explained earlier. However, OFDMA cellular no ACS fixed mod experiences a much faster drop of total data rate offered. This can be explained by the lack of adaptive modulation, which is key when a single user traffic demand is high. This is because high traffic demand of a single user means less users served by a cell and thus lower gain from multiuser diversity. Therefore in such a situation adaptive allocation of subcarriers to users is not sufficient to cope with the traffic demands. This can also be seen in Figure 7.16(b), where blocking probability of OFDMA cellular no ACS fixed mod degrades fast, while for OFDMA cellular and OFDMA cellular no ACS the degradation is moderate. An interesting phenomenon is a high blocking probability for $R = 1$ Mbit/s, which can also be explained by the discrete granularity of the offered resources.

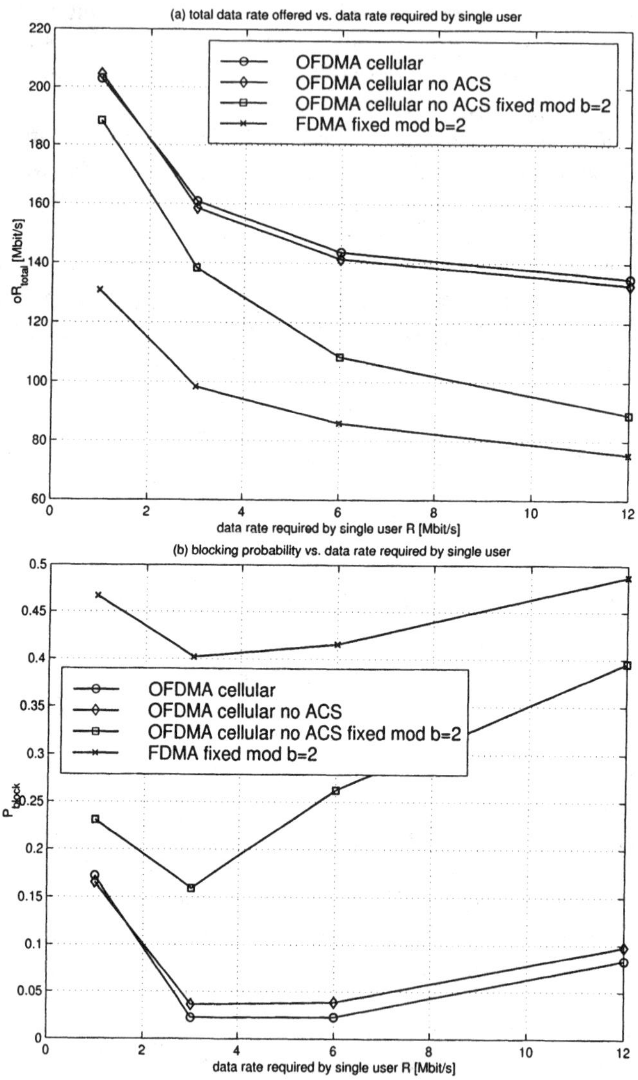

Figure 7.16 Influence of the data rates required by a single user $R = R_u$. The other system parameters are $N = 16$, $\tau_{rms} = 264$ ns, $\alpha = 4.4$, σ^2_{shadow} = 18 dB.

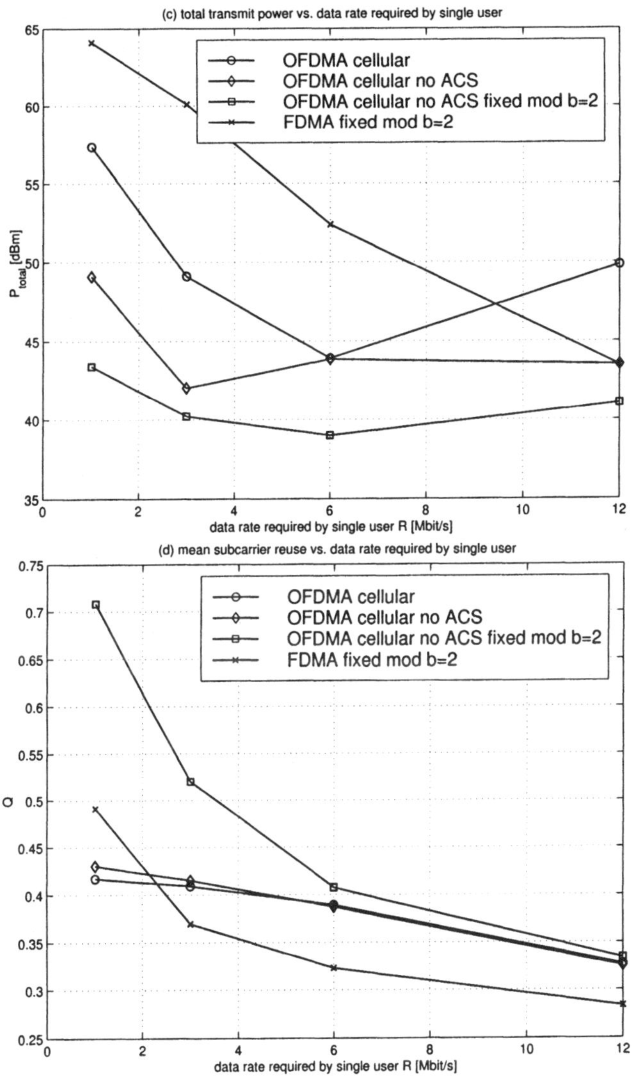

Figure 7.16 (cont.)

The lowest transmit power is required by OFDMA cellular no ACS fixed mod as can be seen in Figure 7.16(c). This is because this scheme is the only one from the OFDMA-family that has the highest blocking probability (thus the highest number of rejected users). Also for OFDMA cellular no ACS fixed mod, the subcarriers' reuse is the highest, as shown in Figure 7.16(d). The reason is the same as in the previous subsection; OFDMA cellular no ACS fixed mod satisfies users' traffic requirements by allocating more subcarriers because it cannot adapt modulation levels. For all the considered schemes, the subcarriers' reuse decreases with higher data rate required by a single user. Again, this is due to lower gain from multiuser diversity, since less users can be served by a cell and the algorithm has less options for adaptive allocation of subcarriers to users.

7.5.5 Influence of path loss

In this subsection, performance evaluation of the considered schemes is presented for various values of path loss exponent, namely: $\alpha = \{2, 3, 4, 5\}$. The results shown in Figure 7.17 (a)-(d) indicate that all the simulated schemes are influenced by the path loss exponent in the same manner. For low α, shielding from CCI is low and therefore the total data rate offered is low, blocking probability is high, and subcarriers' reuse is low. The total transmit power is low for low α because the signal is less attenuated and less transmit power has to be used. The performance measures, except for the total transmit power, improve with higher path loss exponent α.

7.5.6 Influence of shadow fading variance

The performance comparison of the considered schemes for various variances due to shadowing $\sigma^2_{shadow} = \{10, 14, 18, 22\}$ dB, is shown in Figure 7.18 (a)-(d). As can be observed, the shadowing variance does not influence (on average) the performance of the considered systems. It affects all subcarriers belonging to a user in the same way. Therefore it does not add to channel variability of a certain user in the frequency domain (in a way τ_{rms} does).

7.6 SUMMARY

In this chapter, we have considered the problem of radio resource allocation for the downlink cellular OFDMA system with inter-cell reuse of subcarriers. This

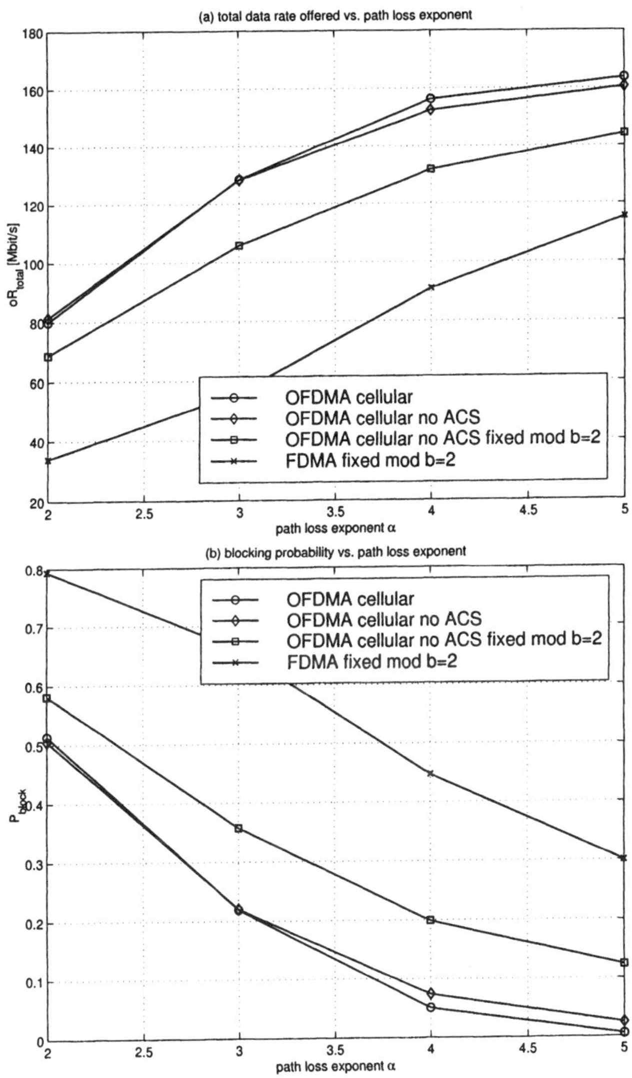

Figure 7.17 Influence of the path loss exponent α. The other system parameters are $N = 16$, $\tau_{rms} = 264$ ns, $R = 3$ Mbit/s, σ^2_{shadow} = 18 dB.

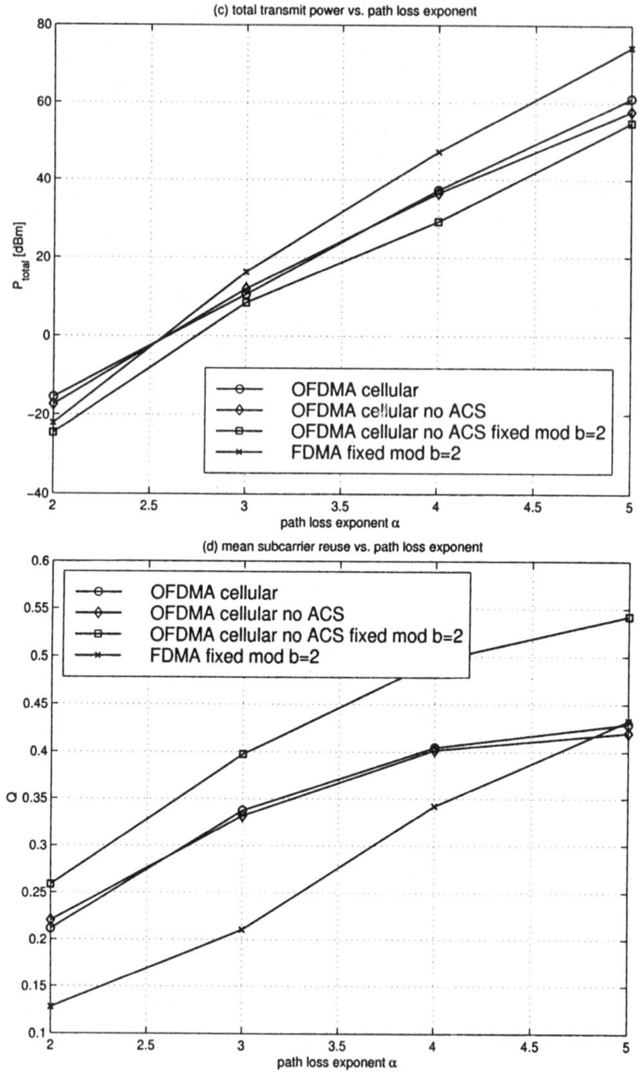

Figure 7.17 (cont.)

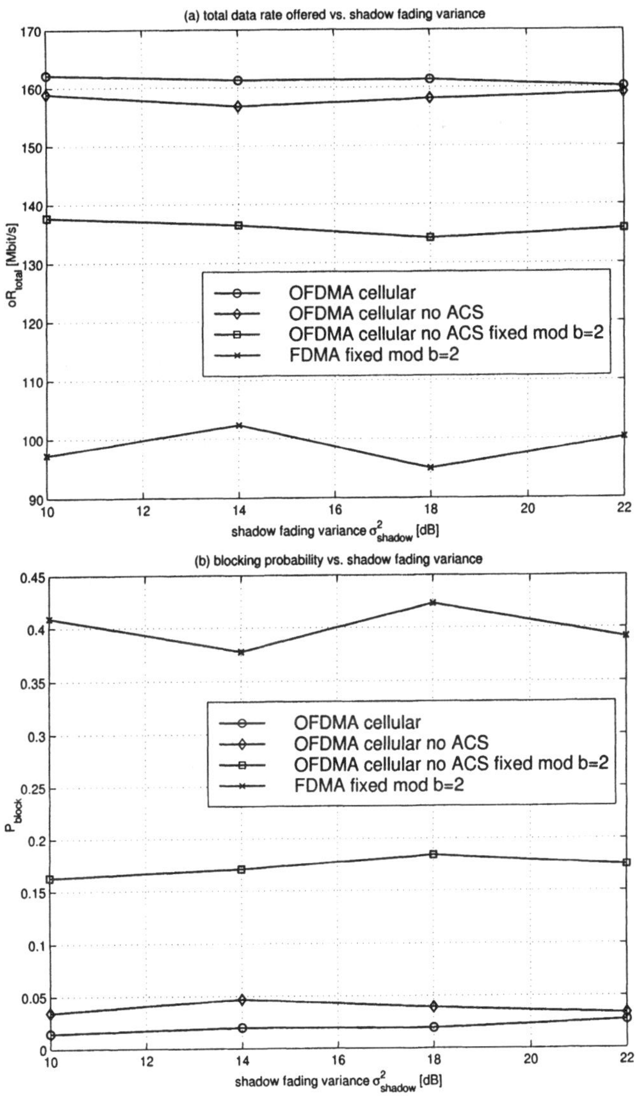

Figure 7.18 Influence of the shadow fading variance σ^2_{shadow}. The other system parameters are $N = 16$, $\tau_{rms} = 264$ ns, $R = 3$ Mbit/s, $\alpha = 4.4$.

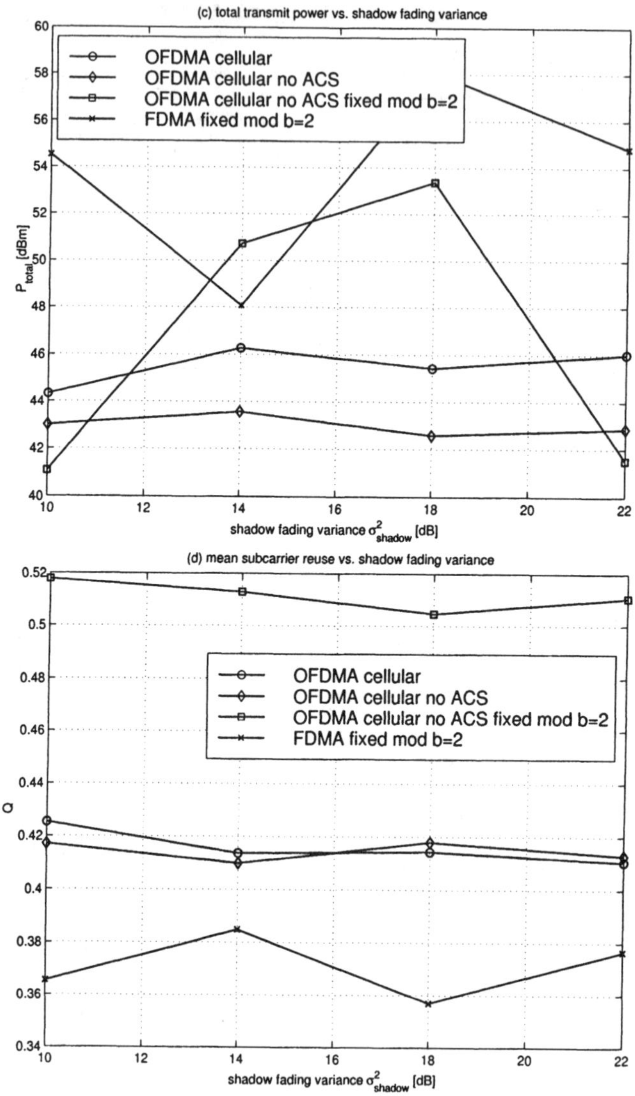

Figure 7.18 (cont.)

includes assignment of users to base stations, subcarriers to users, bit loading, and power control. The cellular scenario poses particular challenges to the wireless system design due to the presence of cochannel interference limiting the system performance.

A set of algorithms was described that solves the resource allocation problem in three steps. The main assumptions taken during the design of the RRM algorithms include support and maintenance of the required QoS, minimization of the transmit power, and application of a reuse factor of one. Such assumptions make the proposed methods suitable for multimedia communications, where, for example, support of various QoS profiles is necessary. In addition, the network is self-organized in frequency reuse, and thus frequency planning is obsolete.

The performance of the presented set of RRM algorithms has been evaluated by simulations. The results indicate superior performance of the described schemes (OFDMA cellular) over a conventional method based on FDMA with power control. At a blocking probability of 0.02 the total data rate achieved with the introduced set of algorithms is about 12 times higher than for FDMA. This gives, on average, a data rate per cell of ~6.58 Mbit/s (~125 Mbit/s divided over 19 cells), with peak rates per cell up to 42 Mbit/s (all subcarriers used with 64-QAM modulation) within a bandwidth of 7 MHz. When compared to the optimal solution obtained by exhaustive search over all possible assignments, the average performance degradation expressed in total transmit power of OFDMA cellular equals ~4.9 dBm, for a system scenario limited to $U = 4$, $N = 4$ $K = 2$, and $BW = 5$ MHz due to excessive computation time. This is a good result since the degradation of FDMA equals ~17.9 dBm. An even smaller difference is expected for OFDMA cellular for systems with higher U, N, and K.

Moreover, it has been observed that the performance of the OFDMA cellular improves with increasing frequency selectivity, which is disastrous for fixed FDMA.

The research area of RRM for cellular OFDMA systems is relatively unexplored in the literature. We believe that the methods described in this chapter may become one of the future enhancements to BWA standards such as IEEE 802.16.

References

[1] Holma, H., and Toskala, A., (eds.), *WCDMA for UMTS : Radio Access for Third Generation Mobile Communications*, Wiley, 2004.

[2] Halonen, T., Romero, J., and Melero, J., (eds.), *GSM, GPRS and EDGE Performance : Evolution Towards 3G/UMTS*, Wiley, 2003.

[3] Lee, W.C.Y., *Mobile Cellular Telecommunications Systems,* McGraw-Hill Book Company, 1989.

[4] Lawrey, E. P., *Adaptive Techniques for Multiser OFDM,* James Cook University, Ph.D. Dissertation, 2001.

[5] Chuang, J.C.I., and Sollenberger, N.R., "Spectrum Resource Allocation for Wireless Packet Access with Application to Advance Cellular Internet Service," *IEEE Journal on Selected Areas in Communications,* August 1998, volume 16, pp. 820-829.

[6] Katzela, I., and Naghshineh, M., "Channel Assignment Schemes for Cellular Mobile Telecommunication Systems: A Comprehensive Survey," *IEEE Personal Communications,* June 1996, pp. 10-31.

[7] Lei, Z., and Slimane, S.B., "A Multicarrier Allocation (MCA) Scheme for Variable-Rate 3G Wireless Systems," *IEEE Communications Magazine,* October 2000, pp. 86-91.

[8] Yih, C., and Geraniotis, E., "Adaptive Modulation, Power Allocation and Control for OFDM Wireless Networks," *Proc. IEEE PIMRC,* 2000.

[9] Zhang, Y., and Letaief, K.B., "Multiuser Subcarrier and Bit Allocation along with Adaptive Cell Selection for OFDM Transmission," *Proc. IEEE ICC,* pp. 861-865, 2002.

[10] Rhee, W., and Cioffi, J.M., "Increase in Capacity of Multiuser OFDM System Using Dynamic Subchannel Allocation," *Proc. IEEE VTC,* Spring, pp. 1085-1089, 2000, Tokyo, Japan.

[11] Li, J., Kim, H., Lee, Y., and Kim, Y., "A Novel Broadband Wireless OFDMA Scheme for Downlink in Cellular Communications," *Proc. IEEE WCNC,* pp. 1907-1911, 2003.

[12] Yin, H., and Liu, H., "An Efficient Multiuser Loading Algorithm for OFDM-based Broadband Wireless Systems," *Proc. IEEE Globecom,* pp. 103-107, 2000.

[13] Kivanc, D., and Liu, H., "Subcarrier Allocation and Power Control for OFDMA," *Proc. 34th Asilomar Conference on Signals, Systems and Computers,* pp. 147-151, November 2000.

[14] Pfletschinger, S., Munz, G., and Speidel, J., "Efficient Subcarrier Allocation for Multiple Access in OFDM Systems," *Proc. Int'l OFDM Workshop,* 2002, Hamburg, Germany.

[15] Yu, W., Ginis, G., and Cioffi, J.M., "An Adaptive Multiuser Power Control Algorithms for VDSL," *Proc. IEEE Globecom,* pp. 394-398, 2001.

[16] Cherubini, G., "Optimum Upstream Power Back-Off and Multiuser Detection for VDSL," *Proc. IEEE Globecom,* pp. 375-380, 2001.

[17] Sonalkar, R.V., and Applegate, D., "Shannon Capacity of Frequency-Overlapped Digital Subscriber Loop Channels," *Proc. IEEE ICC,* pp. 1741-1745, 2002.

[18] Lee, J., Sonalkar, R.V., and Cioffi, J.M., "Multi-User Discrete Bit-Loading for DMT-based DSL Systems," *Proc. IEEE Globecom,* pp. 1259-1263, 2002.

[19] Lee, J., Sonalkar, R.V., and Cioffi, J.M., "A Multi-User Rate and Power Control Algorithm for VDSL," *Proc. IEEE Globecom,* pp. 1264-1268, 2002.

[20] Kulkarni, G., and Srivastava, M., "Subcarrier and Bit Allocation Strategies for OFDMA based Wireless Ad Hoc Networks," *Proc. IEEE Globecom,* pp. 92-96, 2002.

[21] Sonalkar, R.V., "Bit- and Power-Allocation Algorithm for Symmetric Operation of DMT-Based DSL Modems," *IEEE Transactions on Communications,* June 2002, volume 50, pp. 902-906.

[22] Chung, S.T., and Cioffi, J.M., "Rate and Power Control in a Two-User Multicarrier Channel with No Coordination: The Optimal Scheme vs. Suboptimal Methods," *Proc. IEEE VTC,* Fall, pp. 1744-1748, 2002.

[23] Pietrzyk, S., and Janssen, G.J.M., "Subcarrier Allocation and Power Control for QoS Provision in the Presence of CCI for the Downlink of Cellular OFDMA Systems," In Proc. IEEE VTC Spring, pp. 2221-2225, April 2003, Jeju, Korea.

[24] Wong, C.Y., Tsui, C.Y., Cheng, R.S., and Letaief, K.B., "A Real-time Sub-carrier Allocation Scheme for Multiple Access Downlink OFDM Transmission," *Proc. IEEE VTC,* Fall, pp. 1124-1128, 1999.

[25] Pietrzyk, S., and Janssen, G.J.M., "Multiuser Subcarrier Allocation for QoS Provision in the OFDMA Systems," *Proc. IEEE VTC,* Fall, pp. 1077-1081, September 2002, Vancouver, Canada.

[26] Pietrzyk, S., Janssen, G.J.M., "Radio Resource Allocation for Cellular Networks Based on OFDMA with QoS Guarantees," In Proc. IEEE Globecom, November-December 2004, Dallas, Texas, USA.

[27] Pietrzyk, S., and Janssen, G.J.M., "Performance Evaluation of Bit Loading, Subcarrier Allocation and Power Control Algorithms for Cellular OFDMA Systems," *Proc. Int'l Conference on Personal and Wireless Communications,* September 2004, Delft, the Netherlands.

[28] Bambos, N., and Pottie, G.J., "Power Control Based Admission Policies in Cellular Radio Networks," *Proc. IEEE Globecom,* pp. 863-867, 1992.

[29] Chen, S.C., Bambos, N., and Pottie, G.J., "Admission Control Schemes for Wireless Communication Networks with Adjustable Transmitter Powers," *Proc. IEEE Infocom,* pp. 21-28, 1994, Toronto, Canada.

[30] Cioffi, J.M., *A Multicarrier Primer,* Amati Communications Corporation and Stanford University, November 1991.

[31] Kuhn, H.W., "The Hungarian method for the assignment problem," *Naval Research Logistics Quarterly* 2, 1955.

[32] Dondeti, V.R., and Emmons, H., "Max-Min Matching Problems with Multiple Assignments," *Journal of Optimization Theory and Applications,* November 1996, volume 91, pp. 491-511.

[33] Wong, C.Y., Cheng, R.S., Letaief, K.B., and Murch, R.D., "Multiuser OFDM with Adaptive Subcarrier, Bit, and Power Allocation," *IEEE Journal on Selected Areas in Communications,* October 1999, volume 17, pp. 1747-1758.

[34] Yu, W., *Competition and Cooperation in Multi-user Communication Environments,* Stanford University, Ph.D. Dissertation, Department of Electrical Engineering, June 2002.

Chapter 8

The future of OFDMA

The main topic of this book is novel radio resource management (RRM) algorithms for BWA systems based on OFDMA. The desired system characteristics and thus requirements imposed on the design of RRM algorithms include efficient use of the available resources, such as RF bandwidth and transmit power, ability to provide high data rates in a harsh multipath radio channel environment, support of various QoS profiles, and, finally, efficient handling of cochannel interference (CCI), since CCI is the basic limitation when spectrum reuse is involved. Three system cases have been investigated:

1. Single-cell OFDMA without spectrum reuse (exclusive subcarrier assignment);
2. Single-cell OFDMA with intra-cell spectrum reuse;
3. Multicell (cellular) OFDMA with inter-cell spectrum reuse.

In this chapter we summarize the solutions and results presented earlier in this book and try to point out the most promising future applications of OFDMA. Then, a review of other technologies improving BWA systems' performance is presented, which is followed by discussing main research and standardization directions. Finally, we conclude the chapter by casting some long-term views on future design of broadband wireless access systems.

8.1 SINGLE-CELL OFDMA WITHOUT SPECTRUM REUSE

Single-cell OFDMA is investigated in Chapter 5 of this book. This area is relatively well explored in the literature. Therefore, Chapter 5 includes extensive review and classification of the most significant algorithms for single-cell OFDMA without spectrum reuse. These algorithms often lay foundations for more complex routines applied in situations when spectrum reuse is allowed. In addition, Chapter 5 describes a combination of OFDMA resource allocation with a negotiation-based QoS model using the QoS parameters: user-individual bit rate and bit error rate. Also, a performance comparison of adaptive subcarrier allocation with and without adaptive modulation is provided. The main conclusions of Chapter 5 are as follows:

- OFDMA is particularly suitable as a means for providing various levels of QoS, thanks to its inherent granularity of the frequency resource.
- Inclusion of a user-individual BER requirement in the QoS model and in the construction of the cost matrix used for resource allocation makes it possible to use exactly as much transmit power as required without resorting to the worst-case system design.
- In general, adaptive modulation provides a better performance when combined with adaptive subcarrier allocation. However, in special cases both adaptive and fixed modulation exhibit comparable performance. This happens when the total data rate required by users is equal to the maximal data rate that the fixed modulation system can offer. Based on this observation, application of fixed modulation is recommended for systems with stable traffic profiles (fixed required total data rate). For all the other cases adaptive modulation will result in a better performance.

8.2 SINGLE-CELL OFDMA WITH INTRA-CELL SPECTRUM REUSE

This topic is investigated in Chapter 6. The main contributions of this chapter include presentation of the algorithms for efficient distribution of radio resources for OFDMA systems, where subcarriers are reused in the downlink of a cell by applying embedded modulation. As stated in [1], schemes known from broadcast systems, such as hierarchical (embedded) modulation, could also be considered for downlink of a wireless access system, since they exhibit a high potential of providing performance gain over exclusive resource allocation. This is in fact confirmed in Chapter 6. In addition to presenting a number of algorithms for resource allocation, their performance is evaluated under various environmental conditions. The main conclusions of Chapter 6 are as follows:

- Embedded (hierarchical) modulation allows to stack multiple signals on a single subcarrier creating in this way an increased number of independent links and thus increasing capacity of the OFDMA system. When L signals are stacked, the system can offer up to L-times more independent links.
- In the OFDMA context, embedded modulation may be treated as a generalization of adaptive modulation, where bits loaded on a single subcarrier may be individually addressed to different users, depending on the underlying channel conditions. This is in contrast to conventional adaptive modulation, where bits loaded on a single subcarrier are addressed to one user only. This extra freedom in resource allocation makes embedded modulation a more power-efficient scheme than conventional adaptive modulation.
- Higher frequency selectivity due to multipath or more users present in the system improves the performance of the OFDMA with embedded modulation. This is because the adaptive subcarrier allocation algorithm has more options to choose from and can benefit from independent fading patterns of different users (i.e., multiuser diversity). Such conditions are usually disastrous for contemporary systems based on a fixed resource allocation.

8.3 CELLULAR OFDMA WITH INTER-CELL SPECTRUM REUSE

This topic is investigated in Chapter 7 of this book. The majority of the contemporary wireless access systems are built as cellular point-to-multipoint systems. A cellular scenario seems also to be the first choice if we think about widespread application of OFDMA as a technique for next generation networks. Though the application of the cellular principle to OFDMA may seem a natural extension of the studies on OFDMA, not much research work is done on this topic. The main deliverables of Chapter 7 include presentation of radio resource allocation algorithms for appointing the serving base station, allocation of subcarriers to users, adaptation of modulation level (bit loading) and finally control of transmit power. One of the main assumptions taken during the design of the described RRM schemes is the need to provide and to *maintain* the required QoS, which is especially difficult in cellular systems. The performance of the described algorithms is also verified under various environmental conditions. The main conclusions of Chapter 7 are as follows:

- Combination of adaptive subcarrier assignment, adaptive modulation and power control handling CCI leads to substantial gains in total transmit power reduction and decrease of blocking probability.

- Cooperative power control, where the transmit power level for a particular subcarrier is determined together with all the cochannel subcarriers, can be applied not only within a network belonging to the same operator, but also between networks of different operators. Such a situation would be especially beneficial in the license-exempt bands typical for WLAN applications and allowed for WiMAX systems, where adaptation to local interference conditions is of key importance. In this way, instead of harmful competition and mutual jamming these systems could cooperate and peacefully coexist.
- The described radio resource management algorithms make a cellular OFDMA network self-organized in frequency reuse. Subcarriers are not fixed to a base station but are dynamically allocated according to local load and channel conditions. In this way an aggressive reuse factor of one is realized. In addition, this makes frequency planning, which is usually a complex task, obsolete.
- Adaptive cell selection (ACS) increases the chance for user admission.
- Higher frequency selectivity (more multipath) improves the performance of cellular OFDMA thanks to increased multiuser diversity, which exposes more options for making good allocation. Such conditions are usually disastrous for the majority of contemporary systems based on fixed resource allocation.
- Path loss has a significant influence on the capacity of a cellular system since it is a natural means of shielding from CCI,
- It is important to match the allowed traffic profiles with the granularity offered by the system (determined by the number of subcarriers and modulation levels) in order to offer just-as-required data rates.

8.4 APPLICATIONS OF OFDMA

OFDMA offers significant advantages as a multiple access scheme for BWA systems. Below we list their most evident examples:

- OFDMA is robust to multipath propagation since it derives from OFDM scheme.
- OFDMA is bandwidth-efficient.
- OFDMA is relatively easy to implement in the digital domain by the use of (I)FFT.

- OFDMA partitions the frequency resource in a number of independent subcarriers. This provides resource granularity, which allows for providing various levels of QoS.
- OFDMA's resource granularity opens up plenty of possibilities for various allocations of frequency resource to multiple users. They can be used either to minimize the transmit power, maximize the data rate, extend the system's coverage, or make a system more robust to difficult radio conditions.
- OFDMA, thanks to its inherent parallelism in the frequency domain, allows for independent per-subcarrier processing. This may include per-subcarrier (or group of subcarriers such as subchannel in IEEE 802.16) adaptation of modulation modes or power control.

We foresee the application of OFDMA mainly to fixed BWA systems since they are aimed for a high number of users, offering high data rates and MAN coverage. Other possible applications include WLANs and broadcast systems; many of them have OFDM in their PHY layers, which would ease the standard evolution. Some standards, such as WiMAX, are already equipped with OFDMA; some may need redesign and thus consider OFDMA as a future enhancement (e.g., WiFi or DVB-T). For some other systems, such as successors of UMTS/HSDPA, OFDMA may be a totally new PHY layer. As such, it can become the anticipated new radio access technology for systems beyond IMT-2000 [2]. OFDMA is also considered for beyond 3G mobile communication systems by China TDD Special Work Group [3]. It is also part of the air interface of time-division code-division multiplexing OFDM (TD-CDM-OFDM) [4], which is a follower of time-division synchronous CDMA (TD-SCDMA) being a Chinese 3G system.

8.5 OTHER TECHNOLOGIES IMPROVING BWA SYSTEMS' PERFORMANCE

In this section we provide an overview of technologies other than OFDMA that exhibit high potential for improving BWA systems' performance. Some of them can also be combined together with OFDMA to further strengthen the performance improvements. They can constitute (and in many cases they already constitute) prospective enhancements for future releases of BWA standards (such as IEEE 802.16). We shall start with briefly discussing a role of already "classical" schemes such as adaptive modulation and coding (AMC) or power control before going into more advanced solutions such as multiple antennas.

AMC can already be considered a "classical" scheme since it is implemented in most of the commercial wireless access systems including EDGE, HSDPA, WiFi, and WiMAX. The basic principle of AMC is to adapt the modulation and coding mode to changing conditions of the radio channel [5]. When the channel is good, higher-order modulation and less protective coding is chosen, while when the channel is bad, more robust modulation and more protective coding is applied. Thanks to that, system capacity can be increased by more efficient use of resources. AMC can exploit the channel variability both in time and frequency domain. Another "classical" scheme is power control. It can compensate signal fluctuations, reduce cochannel interference, and lower power consumption. It is also worth mentioning here hybrid ARQ (H-ARQ), which is already part of EDGE, HSDPA, and WiMAX standards. H-ARQ allows to retransmit corrupted packets, possibly encoding them with different error correction, and to combine all retransmissions at the receiver so that all retransmitted packets are exploited [6]. Such an approach adds to increased link reliability and improves a system's overall data rate.

8.5.1 Multiple antenna systems

Multiple antennas have been considered for some time already as a means for improving wireless system performance such as range and robustness to fading and interference. When combined with the ability to adapt their configuration, these systems are often referred to as smart or adaptive antennas [7]. Their first applications were in beamforming and diversity provision. In the following, we provide an overview of the main applications of adaptive antennas.

Beamforming can be understood as shaping the antenna radiation pattern in such a way that the antenna gain is maximized in the desired direction. The created beam is formed by appropriate combining of signals coming from different antenna elements, hence its name [8, 9]. Beamforming can improve the range of a wireless system by creating a highly directional antenna [10]. This feature can be well applied to fixed BWA systems requiring LOS (such as LMDS), where LOS is available [8]. It is also possible that beamforming is used for cochannel interference cancellation. This is realized by pointing the antenna pattern null in the direction of cochannel interferer [7]. Because of its ability to filter signals coming from different directions (angles), beamforming is often referred to as spatial filtering.

Another early application of multiple antennas is diversity provision. This is probably the most-often used antenna technique in contemporary commercial wireless access systems (e.g., GSM, UMTS). Antenna diversity reduces the negative effect of fading. The principle is simple: a signal that is faded on one antenna is not likely to be faded on another antenna, assuming that sufficient spatial separation between antennas is provided [7]. Antenna diversity can be applied both at the transmitter (transmit diversity) and the receiver (receive diversity) and is most often used at the BS.

Closely related to beamforming is space division multiple access (SDMA) [11]. In SDMA, communication with different users is realized with the use of independent antenna beams. Thanks to that, users are separated in the space domain so separating them in frequency or time domains is not necessary. Therefore, available spectrum can be reused much more aggressively, which increases the system's capacity. SDMA is also referred to as angle reuse (of frequency) [11].

Beamforming, antenna diversity and SDMA usually require the presence of multiple antennas only at the base station. In recently developed multiple-input multiple-output (MIMO) systems, smart antennas are used both at the transmitter and the receiver, thus both at the BS and SS. In general, MIMO can be used to benefit from diversity or to improve the system's capacity. A major subgroup of MIMO systems providing diversity is called space-time coding. In space-time coding gains are realized by encoding data in space and time domains. The two main types of space-time coding are space-time block codes (STBC) and space-time trellis codes (STTC) [12]. STBCs have lower decoding complexity than STTCs and their popular representative is the Alamouti code [13], described in Chapter 1.

Another group of MIMO systems are those boosting the system's capacity. The idea behind these systems is simple. By having multiple antennas and by appropriately constructing the signals, we can form multiple, separate, and distinguishable links thus increasing the system's capacity [7]. These techniques are also referred to as spatial multiplexing (SM) with the example schemes known as BLAST (Bell Laboratories Layered Space-Time) or V-BLAST (Vertical BLAST) [10, 11]. These schemes are processing signals in the following ways: (1) only at the receiver, (2) both at the transmitter and receiver, or (3) only at the transmitter [7]. Thanks to their promising performance, MIMO systems have already become part of BWA standards such as WiMAX or HSDPA.

MIMO systems are particularly attractive when combined with OFDM. In MIMO-OFDM, each antenna branch at the transmitter and receiver has its own OFDM chain, which includes FEC encoder, modulation mapper, MIMO encoder, and OFDM modulator (IFFT) at the transmitter, with reverse functions at the receiver [12]. Different signals originating from multiple transmit antennas are multiplexed in the air and then present at each receive antenna element of the receiver. Thanks to the application of OFDM, encoding can be performed not only in space and time domains but also in the frequency domain. Thus, with the presence of OFDM subcarriers a system is offered a new dimension, namely frequency dimension, for resource allocation and adaptation of modulation modes and power control.

8.5.2 Multiuser detection

The main principle of multiuser detection schemes is to exploit the undesired signals (interference) in the detection of the desired signals [14, 15]. This is in contrast with most of the contemporary systems where interference is treated as useless background noise. By exploiting the structure of interference and detecting it jointly with the desired signals, we can expect gains in improving spectral efficiency and the system's capacity [16].

8.5.3 Multicarrier CDMA

An interesting combination of OFDM and code division multiple access (CDMA) has been developed in early nineties. This new family of multiple access schemes is referred to as multicarrier CDMA (MC-CDMA) [17]. In MC-CDMA, spectrum spreading is applied before modulating OFDM subcarriers and can be performed either in frequency or time domain. Depending on that, the following members of the MC-CDMA family can be identified: MC-CDMA with frequency domain spreading, multicarrier direct sequence CDMA (DS-CDMA) with time domain spreading, and multitone CDMA (MT-CDMA), where spreading with long codes is also performed in time domain. According to [18], the foreseen application of MC-CDMA schemes is in the downlink and their major advantage over DS-CDMA is the ability to effectively combine received signals scattered in the frequency domain. A combination of OFDM and CDMA referred to as time-division code-division multiplexing OFDM (TD-CDM-OFDM) has been considered as an evolution of the TD-SCDMA [4].

8.6 FUTURE RESEARCH AND STANDARDIZATION DIRECTIONS

Clearly, future wireless access systems will have to support high data rates and high numbers of users, provide various levels of QoS, and use the available resources efficiently. Taking into account mobile subscribers only, their number increased from around 300 million in 1997 to around 800 million in 2001 and is expected to be as high as 1,700 million in 2010 [2]. These requirements form a framework within which the future standardization and research directions will have to operate. In the following, after discussing the prospective future research and standardization directions applicable to wireless access systems in general, we shall outline research activities needed to further improve the resource allocation schemes for OFDMA that are addressed in this book.

8.6.1 Future directions for standard systems

We shall start with presenting major conclusions of the International Telecommunication Union (ITU) on future development of IMT-2000 and systems beyond IMT-2000 [2, 19]. Although they explicitly address mobile systems, key conclusions apply also to fixed BWA systems. It is envisaged that new radio interfaces need to be developed to address the increased user expectations and traffic loads. This will include design of new modulation schemes, multiple access methods, adaptive radio interfaces, and antenna technologies. Efficient spectrum utilization is another hot topic, which can be addressed with new schemes for dynamic channel assignment, spectrum sharing, and novel types of cell structures. It is anticipated that there will be a number of different access networks available, each specialized for a given purpose [20]. A user will be able to utilize multiple services provided by these multiple access networks and the key issues will be the interoperability between the systems and provision of seamless service.

For WLANs, improvements are anticipated in providing more intelligence to PHY and MAC layers. Current implementations of WLANs can be considered 1G WLAN systems. Although they are equipped with schemes such as AMC, they lack many of the functionalities known from the GSM world. It is expected that functionalities such as dynamic channel assignment, power control, or load sharing would be able to improve the capacity of WLAN networks, their data rates, service availability, and at the same time, ease the networks' deployment [21]. As such, they will become major directions for future enhancement to WLAN standards.

In the WiMAX family, improvements are expected in adding mobility to the originally fixed system [22]. This is done in release IEEE 802.16e. Also, further releases will incorporate the cognitive radio functionality. In brief, this is the ability of the access network to smoothly adapt to current radio interference situations. Thus, we can expect that future cognitive radio systems will be interoperable and cooperative.

Finally, it is worth mentioning that almost all standards, including IEEE 802.16, HSDPA, and Chinese Beyond 3G systems foresee the application of multiple antennas as one of the key performance boosters [3, 4]

8.6.2 Future directions for OFDMA RRM techniques addressed in this book

In this subsection we shall outline the key research directions that should be taken to further improve the radio resource allocation algorithms presented in this book. We observe that they have already started to become hot research topics.

Radio resource allocation methods presented in this book require accurate knowledge of the channel state information (CSI). It is therefore important to evaluate the performance of these methods under the condition of nonperfect CSI and/or to redesign the methods in order to make the system robust to partially known CSI.

The considered allocation methods are centrally controlled. It would be interesting to redesign them in order to distribute the control so that individual users could make part or all of the decisions about resource allocation. Such an approach will decrease the required signaling load.

The focus point of the work recorded in this book considers the downlink of a cellular system. However, the uplink is also of interest, where signals corresponding to different users are independently generated. As well, other options for network organization can be considered such as ad hoc or mesh networks, where links are symmetrical (i.e., there is no distinction between the uplink and the downlink) and resource management needs to be distributed. This makes the ad hoc and mesh networks, as well as the uplink for cellular systems, challenging future research areas.

Obviously, we shall expect further performance gains when combining the presented methods with the application of multiple antennas.

Finally, the performance gains observed for the presented adaptive resource allocation methods come with the price of increased computational and signaling load. This price needs closer evaluation. Especially, time required for algorithms' execution should be compared to the channel coherence time.

8.7 CONCLUDING REMARKS

In future BWA systems, a trade-off should be found between performance gain associated with adaptivity and performance loss stemming from system complexity. Such future BWA systems should be able to provide and to maintain simultaneously various QoS profiles required by multiple users. Obviously, they should use the available resources wisely, which translates to the minimal consumption of transmit power and efficient use of the available bandwidth. The need to support multiple users should be exploited as an opportunity (as treated in this book) rather than as a threat. Interference created by the presence of multiple users should be managed in a cooperative way, what is believed to be a successful approach in realizing a future BWA or Beyond-3G wireless access system.

Since OFDM/OFDMA, after losing the battle for 3G, appears to be one of the most promising components for 4G (which is already becoming visible in the recent standardization work [22–24]) it is believed that the resource management methods presented in this book will open new ways of dealing with scarce radio resources. Support and maintenance of QoS, minimization of transmit power, bandwidth efficiency, and cooperative interference handling are the key features of future wireless communication systems.

References

[1] Zhou, S., Li, Y., Zhao, M., Xu, X., Wang, J., and Yao, Y., "Novel Techniques to Improve Downlink Multiple Access Capacity for Beyond 3G," *IEEE Communications Magazine,* January 2005, pp. 61-69.

[2] ITU-R PDNR WP8F *Vision framework and overall objectives of the future development of IMT-2000 and systems beyond IMT-2000,* 2002.

[3] Zhang, P., Tao, X., Zhang, J., Wang, Y., Li, L., and Wang, Y., "A Vision from the Future: Beyond 3G TDD," *IEEE Communications Magazine,* January 2005, pp. 38-44.

[4] Zheng, K., Huang, L., Wang, W., and Yang, G., "TD-CDM-OFDM: Evolution of TD-SDCMA Toward 4G," *IEEE Communications Magazine,* January 2005, pp. 45-52.

[5] Nanda, S., Balachandran, K., and Kumar, S., "Adaptation Techniqes in Wireless Packet Data Services," *IEEE Communications Magazine,* January 2000, pp. 54-64.

[6] Ghosh, A., Wolter, D.R., Andrews, J.G., and Chen, R., "Broadband Wireless Access with WiMAX/802.16: Current Performance Benchmarks and Future Potential," *IEEE Communications Magazine,* February 2005, pp. 129-136.

[7] Murch, R.D., and Letaief, K.B., "Antenna Systems for Broadband Wireless Access," *IEEE Communications Magazine,* April 2002, pp. 76-83.

[8] Blostein, S.D., and Leib, H., "Multiple Antenna Systems: Their Role and Impact in Future Wireless Access," *IEEE Communications Magazine,* July 2003, pp. 94-101.

[9] Dakdouki, A.S, Banket, V., Mykhalov, N.K., and Skopa, A.A., "Downlink Processing Algorithms for Multi-Antenna Wireless Communications," *IEEE Communications Magazine,* January 2005, pp. 122-127.

[10] Parker, S., Sandell, M., Yee, M.S, Sun, Y., Ismail, M., Strauch, P., and McGehan, J., "Space-Time Codes for Future WLANs: Principles, Practice, and Performance," *IEEE Communications Magazine,* December 2004, pp. 96-103.

[11] Sheikh, K., Gesbert, D., Gore, D., and Paulraj, A., "Smart Antennas for Broadband Wireless Access Networks," *IEEE Communications Magazine,* November 1999, pp. 100-105.

[12] Yang, H., "A Road to Future Broadband Wireless Access: MIMO-OFDM-Bassed Air Interface," *IEEE Communications Magazine,* January 2005, pp. 53-60.

[13] Alamouti, S.M., "A Simple Transmit Diversity Technique for Wireless Communications," *IEEE JSAC,* volume 16, no. 8, October 1998, pp. 1451-1458.

[14] Pottie, G.J., "System Design Choices in Personal Communications," *IEEE Personal Communications,* October 1995, volume 2, pp. 50-67.

[15] Verdu, S. *Multi-user Detection,* Cambridge University Press, New York, 1998.

[16] Verdu, S., "Wireless Bandwidth in the Making," IEEE Communications Magazine, July 2000, pp. 53-58.

[17] Hara, S., and Prasad R., "Overview of Multicarrier CDMA," *IEEE Communications Magazine,* December 1997, pp. 126-133.

[18] van Nee, R., and Prasad, R. *OFDM Wireless Multimedia Communciations,* Artech House, 2000.

[19] Kim, Y., Jeong, B.J., Chung, J., Hwang, C., Ryu, J.S., Kim, K., and Kim, Y.K., "Beyond 3G: Vision, Requirements, and Enabling Technologies," *IEEE Communications Magazine,* March 2003, pp. 120-124.

[20] Ohmori, S. , Yamao, Y., and Nakajima, N., "The Future Generations of Mobile Communications Based on Broadband Access Technologies," *IEEE Communications Magazine,* December 2000, pp. 134-142.

[21] Hills, A., and Friday, B., "Radio Resource Management in Wireless LANs," *IEEE Communications Magazine,* December 2004, pp. S9-S15.

[22] IEEE, *Part 16: Air Interface for Fixed Broadband Wireless Access Systems,* IEEE Std 802.16-2004, October 2004.

[23] ETSI, *ETSI TS 101 475. Broadband Radio Access Networks (BRAN); HIPERLAN Type 2; Physical (PHY) layer,* December 2001.

[24] ETSI, *ETSI TS 101 761-2. Broadband Radio Access Networks (BRAN); HIPERLAN Type 2; Data Link Control (DLC) Layer; Part 2: Radio Link Control (RLC) sublayer,* January 2002.

Appendix A

Hungarian algorithm

In this appendix, the basic idea of the so-called Hungarian algorithm is presented. The Hungarian algorithm is known to provide an optimal solution to the *assignment problem*.

The assignment problem can be explained by the following real-life example. Assume a set of workers $u = 1, \ldots, N$ and a set of tasks $n = 1, \ldots, N$. The number of workers and the number of tasks are the same. The cost of each worker u performing the task n is expressed by $[Q_{u,n}]$. Each worker can be assigned only one task and each task can be assigned to only one worker. The goal is to assign tasks to workers (or workers to tasks) in such a way that the total cost is as low as possible. This can be formally expressed as

$$\min \sum_{u=1}^{N} \sum_{n=1}^{N} C_{u,n} Q_{u,n} \tag{A.1}$$

subject to

$$\sum_{u=1}^{N} C_{u,n} = 1; n = 1, \ldots, N, \tag{A.2}$$

$$\sum_{n=1}^{N} C_{u,n} = 1; u = 1, \ldots, N, \tag{A.3}$$

$$C_{u,n} \in \{0, 1\}; u = 1, \ldots, N; n = 1, \ldots, N. \tag{A.4}$$

where $C_{u,n} = 1$ indicates that user u is assigned task n, while $C_{u,n} = 0$ indicates that user u is not assigned task n.

The complete description of the Hungarian algorithm solving the assignment problem (A.1)-(A.4) optimally was presented by Kuhn in [1] together with the underlying theory. However, the origins of the method come from the earlier papers by Konig [2] and Egervary [3], who were Hungarian scientists and from whom the algorithm derived its name. In the following, the principles of operation of the Hungarian algorithm are presented, which are based on [4,5].

Let us assume that the cost matrix $Q_{u,n}$ is nonnegative (i.e., $Q_{u,n} \geq 0; u = 1, \ldots, N; n = 1, \ldots, N$). In such a case, assignment of value 0, that is the assignment of a particular worker $\overline{u}$ to a particular task $\overline{n}$ for which $Q_{\overline{u},\overline{n}} = 0$, is optimal. If it is possible to assign all workers $u = 1, \ldots, N$ and tasks $n = 1, \ldots, N$ (i.e., to determine the matrix $[C_{u,n}]$) in such a way that $Q_{u,n} = 0$ so that

$$\sum_{u=1}^{N}\sum_{n=1}^{N} C_{u,n}Q_{u,n} = 0 \tag{A.5}$$

then such assignment $[C_{u,n}]$ referred to as the *all-zero* assignment is optimal and solves the assignment problem. The Hungarian algorithm aims at iteratively generating zeros in subsequent equivalent versions of the original cost matrix, which leads to the all-zero assignment. In the following, a simple example of the cost matrix is used as presented below

$$Q_{u,n} = \begin{bmatrix} 8 & 10 & 7 \\ 10 & 11 & 8 \\ 5 & 6 & 7 \end{bmatrix}. \tag{A.6}$$

The Hungarian algorithm operates as follows. In the first step, for each row row_u of $[Q_{u,n}]$ the minimal element $x_u = \min(row_u)$ is found. Then, this element is subtracted from each element of a given row. In the considered example, $x_1 = 7, x_2 = 8$ and $x_3 = 5$. In this way matrix $[Q_{u,n}(1)]$ is created. Then, using matrix $[Q_{u,n}(1)]$ the same is done for each column col_n. Thus we find the minimal elements $y_n = \min(col_n), n = 1, \ldots, N$ and subtract them from elements of the corresponding columns, which results in matrix $[Q_{u,n}(2)]$. In the considered example, $y_1 = 0, y_2 = 1$ and $y_3 = 0$. The original problem does not change after these manipulations since the value of each assignment goes down by the same amount [4]. Matrices $[Q_{u,n}(1)]$ and $[Q_{u,n}(2)]$ corresponding to the example cost matrix are presented below

$$Q_{u,n}(1) = \begin{bmatrix} 1 & 3 & 0 \\ 2 & 3 & 0 \\ 0 & 1 & 2 \end{bmatrix}, \quad Q_{u,n}(2) = \begin{bmatrix} 1 & 2 & 0 \\ 2 & 2 & 0 \\ 0 & 0 & 2 \end{bmatrix}. \tag{A.7}$$

The second step of the algorithm is referred to as the *zero cover*. Let us name a *line* as a row or a column, which *covers* all the elements belonging to it. The idea is to find such set of lines that covers all zeros in the cost matrix $[Q_{u,n}(2)]$ with the minimal possible number of lines indicated by λ_{min}. In the considered example, $\lambda_{min} = 2$, since in order to cover all zeros in the cost matrix $[Q_{u,n}(2)]$, one line goes over row row_3 and one line over column col_3. If $\lambda_{min} = N$ (which is not the case of our example, where $N = 3$) then this means that we have an all-zero cover, which determines the optimal all-zero assignment so the algorithm terminates. Otherwise the algorithm goes to the next step.

The lines determine two sets of entries. The first set are those entries that are not covered by any line and they are referred to as *uncovered*. In the considered example these are the entries: $(1,1), (1,2), (2,1), (2,2)$. The second set includes those entries that are covered by two lines and they are referred to as $doublecovered$. In the considered example entry $(3,3)$ is $doublecovered$. In the third step, the smallest $uncovered$ entry is found, $z = \min(Q_{u,n}), (u,n) \in uncovered$. In our example case $z = 1$, which corresponds to entry $(1,1)$. Then, z is subtracted from each $uncovered$ entry, so $Q_{u,n} = Q_{u,n} - z$ if $(u,n) \in uncovered$, and is added to each $doublecovered$ entry, so $Q_{u,n} = Q_{u,n} + z$ if $(u,n) \in doublecovered$. This forms the next version of the cost matrix $[Q_{u,n}(3)]$. Now, again zero cover is applied and it is checked if $\lambda_{min} = N$ (i.e., if all-zero cover is available). If so, the optimal assignment matrix $[C_{u,n}]$ is found by setting $C_{\overline{u},\overline{n}} = 1$ for such $\overline{u}$ and $\overline{n}$ for which $Q_{\overline{u},\overline{n}}(3) = 0$, removing row $row_{\overline{u}}$ and column $col_{\overline{n}}$ from $[Q_{u,n}(3)]$ until all zeros are picked up. If, however, $\lambda_{min} < N$, then the last (third) step should be repeated until $\lambda_{min} = N$. In the considered example $\lambda_{min} = 3$ so no further zero search is necessary. The third version of the cost matrix $[Q_{u,n}(3)]$ and the final assignment matrix $[C_{u,n}]$ are given below

$$Q_{u,n}(3) = \begin{bmatrix} 0 & 1 & 0 \\ 1 & 1 & 0 \\ 0 & 0 & 3 \end{bmatrix}, \quad C_{u,n} = \begin{bmatrix} 1 & 0 & 0 \\ 0 & 0 & 1 \\ 0 & 1 & 0 \end{bmatrix}. \tag{A.8}$$

Such assignment matrix leads to the total cost, which equals $\sum_u \sum_n C_{u,n} Q_{u,n} = 8 + 6 + 8 = 22$ and which is optimal.

The complete Hungarian algorithm is depicted as Algorithm A.1.

Algorithm A.1: Hungarian algorithm providing optimal solution to the assignment problem.

input : $[Q_{u,n}], u = 1, \ldots, N; n = 1, \ldots, N$
output: $[C_{u,n}]$

begin
  for $u = 1$ **to** N **do**
    $x_u = \min(row_u)$
    $row_u = Q_{u,n} - x, n = 1, \ldots, N$
    update $Q_{u,n}$ with row_u
  for $n = 1$ **to** N **do**
    $y_n = \min(col_n)$
    $col_n = Q_{u,n} - y, u = 1, \ldots, N$
    update $Q_{u,n}$ with col_u
  do *zero cover*: determine $\lambda_{min}, uncovered, doublecovered$
  while $\lambda_{min} < N$ **do**
    $z = \min(Q_{u,n}), (u, n) \in uncovered$
    $Q_{u,n} = Q_{u,n} - z$ if $(u, n) \in uncovered$
    $Q_{u,n} = Q_{u,n} + z$ if $(u, n) \in doublecovered$
    determine $\lambda_{min}, uncovered, doublecovered$
  find *all-zero* assignment in $[Q_{u,n}]$ and indicate it with 1-s in the assignment matrix $[C_{u,n}]$
end

References

[1] Kuhn, H.W., "The Hungarian Method for the Assignment Problem," *Naval Research Logistics Quarterly* 2, 1955.

[2] Konig, D., "Uber Graphen und ihre Anwendung auf Determinantentheorie und Mangenlehre," *Math. Ann.*, 1916, volume 77, pp. 453-465.

[3] Egervary, J., "Matrixok kombinatorius tulajdonsagairol," *Mat. Fiz. Lapok,* 1931, pp. 16-28.

[4] http://www.orie.cornell.edu, *The transportation and assignment problems. Engineering Applications of OR. Lecture handouts,* School of Operations Research and Industrial Engineering, College of Engineering, Cornell University, 1998.

[5] Jonker, R., and Volgenant, T., "Improving the Hungarian Assignment Algorithm," *Operations Research Letters,* October 1986, pp. 171-175.

Appendix B

Review of power control schemes

In this appendix, a review of the literature on power control is presented.

Power control is one of the key functionalities widely implemented in contemporary communication systems. It can be used to:

- Reduce the chance of receiver saturation due to a closely located BS / user terminal;
- Reduce CCI and thus enable higher frequency reuse and capacity;
- Reduce adjacent channel interference (ACI);
- Reduce power consumption of the mobile station, which not only adds to a prolonged battery lifetime but may even be considered as an enabler for some power-hungry services (e.g., high-definition video);
- Compensate for signal strength fluctuations (e.g., due to fading) in order to protect radio link quality.

Radio link quality is proportional to SIR, which is directly proportional to the power of the desired signal and inversely proportional to the sum of the powers of all undesired interfering signals from other users operating at the same frequency [1]. In order to satisfy the link quality demand of a user we cannot simply increase its transmit power, since this would lead to quality degradation for the other cochannel users. Instead, a kind of "equilibrium" should be found, which is a situation where transmit powers of all cochannel users are set in such a way that the resulting SIR is acceptable for each user. Various algorithms attempted to address the problem of finding this "equilibrium." In the following, we briefly go through the findings of the most important papers on power control.

In [2], Zander considered the SIR balancing method, where the goal is to minimize the interference probability, which in effect is maximizing the minimum SIR present in the system. The proposed method is fully centralized and requires the knowledge of all link gains in the system. Therefore it is not really practical for direct implementation, but can rather be considered as an upper bound for the system's performance. In this work, Zander suggested the possibility to reduce complexity by using limited knowledge about link gains rather than full knowledge. This clearly indicates a trade-off between performance and complexity.

In [3], a distributed iterative scheme is also proposed by Zander, where power control is based only on local link gain measurements. However, the measurements must be reported back to the BS so that the BS can decide how to adjust the transmit powers; hence the scheme is not fully distributed. Also, the target SIR is common for all users, which is acceptable for single-service networks, such as voice networks, but is not acceptable for multimedia networks. Moreover, during SIR balancing, existing links may observe their SIR to drop below acceptable level.

In [4], a fully decentralized iterative scheme is proposed by Foschini and Miljanic, where each user adjusts its power to the target SIR (based on whether the received SIR is below or above the target SIR) as if the other cochannel users were not going to change their powers. After several iteration steps, the algorithm achieves convergence provided a feasible transmit power vector exists. The problem arises when a feasible transmit power vector does not exist. In such a situation the system becomes unstable (transmit powers go to infinity). Also, during a transient state (before convergence) the SIR of already existing links can be corrupted by a newly arriving link. Also here, the SIR target is assumed to be equal for all users.

Bambos and Pottie in their early paper [5] took a slightly different orientation. They allowed for different users to have different QoS (SIR) targets and aimed explicitly at guaranteeing the QoS of already established links. The construction of a power vector is based on the computationally effective and well understood Gaussian elimination routine. Moreover, power control here is closely related to admission control: if a feasible power vector cannot be formed for a newly arriving call, then the call is blocked. However, this early paper [5] assumes knowledge of all link gains (obtained e.g., by the use of pilots) and thus should rather be treated as the one for finding a performance upper-bound than the one ready for practical implementation.

In [6], a closer relation of power control to admission control is given by Chen, Bambos, and Pottie. The decision about acceptance of a new call is based on the existence of a feasible power vector. It is shown how to reduce the computational complexity of the Gaussian elimination process used to solve the feasibility equations. Also, the issue of reducing computational and signaling load is addressed by proposing a scheme where all cells are divided into categories (those that are close enough to be affected by power control in a given cell and those that are far enough to not be affected) and power control is based on local link gain measurements. This idea reduces the computational and signaling load (in a way similar to [3]) but is not fully distributed.

A fully distributed scheme with individual QoS for users and with QoS guarantees for existing links is presented in [7]. The proposed algorithm is similar to SIR balancing, but thanks to the application of a safety margin (added to the required SIR) and a fixed power increment at each iteration step, it is able to protect the already active links, which is a key feature. The potential drawbacks include a slower convergence achieved by this scheme (due to fixed power increments) and a too stringent target SIR.

A nice tutorial review of power control strategies is given in [8]. All previously mentioned ideas are raised. In addition, the idea of channel probing is considered. Also, the problem of channel selection is considered, since usually we can compose a feasible link in several ways depending on which channel we choose. The issues of adaptation are considered and it is suggested that power control should rather react to average SIR trends than to its random fluctuations. An interesting dilemma is stated: is it better to transmit, when the interference in the air is high or maybe it is better to wait, keep the data in the buffer for some time and transmit when the interference is relieved.

In this book, in Chapter 7 considering a cellular OFDMA system, power control is based on the methods introduced by Bambos [5–8]. This is due to the following reasons:

- It allows for different SIR targets for different users;
- It protects QoS of ongoing links;
- It can be implemented as a global, local, or fully distributed scheme;
- It is part of the admission control;
- It is easy to implement.

References

[1] Gibson, J.D., (ed.), *The Mobile Communications Handbook*, CRC Press, 1999.

[2] Zander, J., "Performance of Optimum Transmitter Power Control in Cellular Radio System," *IEEE Transactions on Vehicular Technology,* February 1992, volume 41, pp. 57-62.

[3] Zander, J., "Distributed Cochannel Interference Control in Cellular Radio Systems," *IEEE Transactions on Vehicular Technology,* August 1992, volume 41, pp. 305-311.

[4] Foschini, G.J., and Miljanic, Z., "A Simple Distributed Autonomous Power Control Algorithm and its Convergence," *IEEE Transactions on Vehicular Technology,* November 1993, volume 42, pp. 641-646.

[5] Bambos, N., and Pottie, G.J., "Power Control Based Admission Policies in Cellular Radio Networks," *Proc. IEEE Globecom,* pp. 863-867, 1992.

[6] Chen, S.C., Bambos, N., and Pottie, G.J., "Admission Control Schemes for Wireless Communication Networks with Adjustable Transmitter Powers," *Proc. IEEE Infocom,* pp. 21-28, 1994, Toronto.

[7] Bambos, N.D., Chen, S.C., and Pottie, G.J., "Radio Link Admission Algorithms for Wireless Networks with Power Control and Active Link Quality Protection," *Proc. IEEE Infocom,* pp. 97-104, 1995.

[8] Bambos, N., "Toward Power-Sensitive Network Architectures in Wireless Communications: Concepts, Issues, and Design Aspects," *IEEE Personal Communications,* June 1998, pp. 50-59.

List of algorithms

In this appendix a list of all algorithms described in this book is presented. The algorithms are listed in the order of their introduction in the book.

LIST OF ALGORITHMS

List of symbols

In this appendix a list of the most important mathematical symbols used in the book is given.

A	amplitude of the dominant path	
$A_{u,k}$	allocation indicator of user u to cell k	
$\boldsymbol{A}$	user-to-cell allocation matrix	
$\boldsymbol{A}_{temp}$	temporal user to cell allocation matrix	
b	number of bits per symbol	
$back$	background cell index	
b_{max}	maximal number of loaded bits	
b_{min}	minimal number of loaded bits	
$b_{u,n}$	number of bits allocated to user u at subcarrier n	
$b_{u,n,k}$	number of bits allocated to user u at subcarrier n in cell k	
$\boldsymbol{B}$	bit allocation matrix	
B_{coh}	coherence bandwidth	[Hz]
$\boldsymbol{B}_{temp}$	temporal bit allocation matrix	
BW	system bandwidth	[Hz]
$C_{u,n}$	allocation indicator of subcarrier n to user u	
$C_{u,n,k}$	allocation indicator of subcarrier n to user u in cell k	
$C_{u,n,l}$	allocation indicator of waveform (n,l) to user u	
$\boldsymbol{C}$	subcarrier allocation matrix	
$\boldsymbol{C}_{temp}$	temporal subcarrier allocation matrix	
$\underline{C}_{v,n,l}$	allocation indicator of waveform (n,l) to virtual user v	
d	distance	[m]
d_0	reference distance	[m]
$d_{u,n}$	distance between constellation points for user u at subcarrier n	
$d_{u,n,l}$	distance between constellation symbols for waveform (n,l) of user u	

$\underline{d}_{v,n,l}$	distance between constellation symbols for waveform (n,l) of virtual user v	
$\underline{d}'_{v,n,l}$	distance between constellation points for waveform (n,l) of virtual user v	
D	reuse distance	
$\underline{D}_{v,n,l}$	constellation width for waveform (n,l) of virtual user v	
$\underline{\widetilde{D}}_{v,l}$	average (over virtual users) constellation width for waveform $(n,l-1)$	
E_n	post-equalization coefficient for subcarrier n	
f	frequency	[Hz]
f_c	carrier frequency	[Hz]
f_d	Doppler frequency	[Hz]
$[F_j]$	list of preferred cells	
$\overline{G}_c$	average channel power response	
$G_{u,n}$	channel gain of subcarrier n between BS and user u	
$G_{u,n,k}$	channel gain on subcarrier n between BS k and user u	
$\underline{G}_{v,n}$	channel gain of subcarrier n between BS and virtual user v	
$h(\tau)$	time-invariant complex baseband channel impulse response	
$h(\tau;t)$	time-variant complex baseband channel impulse response	
$H_{u,n}$	complex channel frequency response sample of user u on subcarrier n	
$H(f)$	complex frequency response of the channel	
H_n	channel frequency response sample	
i	index of an interfering cochannel BS	
$I_{u,n,k}$	interference power experienced by user u on subcarrier n in cell k	
k	cell (base station) index	
$\overline{k}$	a particular cell index	
K	number of cells	
K_{factor}	Rician factor K	
$K_{max}^{checked}$	maximum number of cells on the preferred list	
l	layer index	
$\overline{l}$	a particular layer index	
L	number of layers	
M	modulation level	
$M_{u,n}$	modulation level for user u at subcarrier n	
$M_{u,n,k}$	modulation level for user u at subcarrier n in cell k	

n	OFDM subcarrier index	
$\overline{n}$	a particular subcarrier index	
N	number of OFDM subcarriers	
oPe_u	bit error rate offered to user u	
$oQoS_u$	QoS profile offered to user u	
oR_{total}	total offered bit rate	[bit/s]
oR_u	bit rate offered to user u	[bit/s]
p	path index in multipath channel model	
$\mathbf{p}(n)$	feasible power vector for subcarrier n	
P	number of paths in multipath channel model	
P_{block}	blocking probability	
Pe_u	bit error rate required by user u	
$prim$	primary cell index	
$PL(d)$	path loss at distance d	
P_{max}	BS maximal transmit power	[W]
P_{total}	BS total transmit power	
$P_{u,n}$	transmit power allocated to user u at subcarrier n	
$P_{u,n}^{min}$	minimal transmit power allocated to user u at subcarrier n	
$P_{u,n,k}$	transmit power allocated to user u at subcarrier n in cell k	
$P_{u,n,k}^{min}$	minimal transmit power allocated to user u at subcarrier n in cell k	
$P_{u,n,l}$	transmit power allocated to waveform (n, l) of user u	
P_{back}^{av}	average transmit power "per-subcarrier" in background cell	[W]
P_{prim}^{av}	average transmit power "per-subcarrier" in primary cell	[W]
$\overline{P}_u$	average (over subcarriers) transmit power of user u	
$\underline{P}_{v,n,l}$	transmit power allocated to waveform (n, l) of virtual user v	
$\underline{P}_{v,n,l}^{min}$	minimal transmit power allocated to waveform (n, l) of virtual user v	
$\boldsymbol{P}_{min}$	minimal transmit power allocation matrix	[W]
$\boldsymbol{P}_{back}^{min}$	minimal transmit power allocation matrix in background cell	[W]
$\boldsymbol{P}_{prim}^{min}$	minimal transmit power allocation matrix in primary cell	[W]
$\boldsymbol{P}_{min}^{temp}$	temporal minimal transmit power allocation matrix	
Q	mean subcarrier reuse	
Q_{back}^{ful}	fulfillment coefficient in background cell	
Q_{prim}^{ful}	fulfillment coefficient in primary cell	
r	cell radius	[m]
$rQoS_u$	QoS profile required by user u	

R	bit rate	[bit/s]
R_u	bit rate required by user u	[bit/s]
R_{total}	total bit rate required by all users	[bit/s]
R_{total}^{max}	maximal total bit rate that can be offered by a system	[bit/s]
R_{ref}	reference system load	[bit/s]
s	OFDM symbol index	
$s(t)$	equivalent low-pass transmitted signal	
s_k^{min}	minimal number of subcarriers allocated already to cell k	
s_u^{max}	maximal number of subcarriers required by user u	
s_u^{min}	minimal number of subcarriers required by user u	
s_u	number of subcarriers per user u	
t	time variable	[s]
T	(OFDM) symbol duration	[s]
T_{coh}	coherence time of the channel	[s]
T_e	effective OFDM symbol time duration	[s]
T_g	duration of cyclic prefix	[s]
$T_{u,n}$	normalized channel gain of user u at subcarrier n	
$T_{u,n,k}$	normalized channel gain of user u at subcarrier n in cell k	
$\overline{T}_u$	average (over subcarriers) normalized channel gain of user u	
$\overline{T}_{u,k}$	average (over subcarriers) normalized channel gain of user u in cell k	
$\underline{T}_{v,n,l}$	normalized channel gain of subcarrier n and layer l of virtual user v	
$\overline{u}$	a particular user index	
U	number of users	
U_{back}	number of users in background cell	
U_{prim}	number of users in primary cell	
v	virtual user index	
V	number of virtual users	
V_l	number of virtual users at layer l	
w	waveform index	
W_u	number of waveforms required by user u	
$x(t)$	transmitted signal	
X_n	transmit signal sample	
$X_{n,s}$	complex data symbol in OFDM symbol s on subcarrier n	
$X_{u,n}$	complex data symbol of user u on subcarrier n	
X_{shadow}	shadowing variable	
$y(t)$	received signal	

y_i	instantaneous received signal amplitude	
Y_n	receive OFDM signal sample	
α	path loss exponent	
β_p	amplitude gain of path p	
Δf	separation between consecutive OFDM subcarriers	[Hz]
$\eta(t)$	AWGN noise	
η_n	noise sample in the OFDM receiver	
$\gamma_{u,n}$	SNR of user u at subcarrier n	
$\gamma_{u,n,k}$	SNIR of user u at subcarrier n in cell k	
Γ	SNR gap	
$\underline{\Gamma}$	SNR gap corresponding to virtual user	
κ	number of cochannel cells	
λ	wavelength	[m]
ν	index to a user assigned to another cell on cochannel subcarrier	
$\Phi_n(t)$	OFDM transmitter waveform	
$\Psi_n(t)$	OFDM receiver waveform	
σ_s^2	average scattered signal power	
σ_{shadow}	standard deviation of shadowing random variable	
σ	standard deviation of AWGN	
τ	delay time	[s]
τ_m	average delay time	[s]
τ_p	delay time of path p	[s]
τ_{rms}	root mean square delay spread (rds)	[s]
Θ_p	phase of path p	
$\Upsilon_{v,n,l}$	normalized channel gain at layer l (dough laying algorithm)	
ϑ	relative velocity between the transmitter and the receiver	[m/s]
$*$	complex conjugate	
$\delta(\cdot)$	Dirac delta impulse function	
$\otimes$	convolution operator	
$I_0(\cdot)$	zero-order Bessel function of the first kind	
$Re\{\cdot\}$	real part of complex variable $\{\cdot\}$	

List of abbreviations

The abbreviations used in this book are listed below in the alphabetical order.

AAS	adaptive antenna system
ACI	adjacent channel interference
ACS	adaptive cell selection
AD	analog-to-digital
ADSL	asymmetrical digital subscriber line
AM	amplitude modulation
ARQ	automatic repeat request
ATM	asynchronous transfer mode
AWGN	additive white Gaussian noise
BE	best effort
BER	bit error rate
BPSK	binary phase shift keying
BS	base station
BWA	broadband wireless access
CBR	constant bit rate
CCI	cochannel interference
CD	compact disc
CDF	cumulative distribution function
CDMA	code division multiple access
C/I	carrier to interference
CID	connection identifier
CoS	cost of service
CP	cyclic prefix
CRC	cyclic redundancy check
CSI	channel state information
CSMA	carrier sense multiple access
CSMA/CD	carrier sense multiple access / collision detection
dB	decibel

dBm	decibel relative to one milliwatt
DA	digital-to-analog
DAB	digital audio broadcasting
DCA	dynamic channel allocation
DFT	discrete Fourier transform
DS	direct sequence
DSP	digital signal processing
DVB	digital video broadcasting
DVB-T	digital video broadcasting-terrestrial
ETSI	European Telecommunications Standards Institute
FDD	frequency division duplex
FDMA	frequency division multiple access
FFT	fast Fourier transform
FH	frequency hopping
FM	frequency modulation
FTP	File Transfer Protocol
GPRS	general packet radio service
GSM	global system for mobile communication
HDSL	high-bit-rate digital subscriber line
HDTV	high-definition television
HSDPA	high speed downlink packet access
HUMAN	high-speed unlicensed metropolitan area network
ICI	intercarrier interference
IDFT	inverse discrete Fourier transform
IEEE	Institute of Electrical and Electronics Engineers
IFFT	inverse fast Fourier transform
IP	Internet Protocol
IR	incremental redundancy
ISI	intersymbol interference
ISM	industrial, scientific, and medical band
LAN	local area network
LC	logical channel
LOS	line-of-sight
MAC	medium access control layer
MAN	metropolitan area network
MBWA	mobile broadband wireless access
MC-CDMA	multicarrier code division multiple access
MCM	multicarrier modulation
MIMO	multiple input multiple output

MS	mobile station
NET	network layer
NEXT	near-end cross talk
NLOS	non-line-of-sight
nrtPS	nonreal-time polling service
OFDM	orthogonal frequency division multiplexing
OFDMA	orthogonal frequency division multiple access
PAP	peak-to-average power problem
PDF	probability density function
PDP	power delay profile
PDU	protocol data unit
PHS	payload header suppression
PHY	physical layer
PMP	point-to-multipoint
PN	pseudonoise
QAM	quadrature amplitude modulation
QoS	quality of service
QPSK	quadrature phase shift keying
rds	root mean square delay spread
RF	radio frequency
RM	resource management layer
RRM	radio resource management/manager
rtPS	real-time polling service
RX	receiver
SAP	service access point
SDU	service data unit
SNR	signal-to-noise ratio
SNIR	signal-to-noise-plus-interference ratio
SS	subscriber station
STC	space time coding
TD-CDM-OFDM	time-division code-division OFDM
TDD	time division duplex
TDMA	time division multiple access
TD-SCDMA	time-division synchronous CDMA
TX	transmitter
UGS	unsolicited grant service
UMTS	universal mobile telecommunication system
UNII	unlicensed national information infrastructure

VBR	variable bit rate
VC	virtual connection
VP	virtual path
WLAN	wireless local area network

About the author

Sławomir Pietrzyk is the founder of Innovative Solutions, a consulting and development company supporting its clients in design and implementation of cutting-edge wireless access systems (http://www.InnovativeSolutions.pl).

Sławomir completed his Ph.D. degree at the Delft University of Technology in the Netherlands in 2005 and his M.Sc. degree (magister inżynier) at the Kielce University of Technology in Poland in 1997. In 2003 he received a postgraduate diploma in management from the Warsaw School of Economics.

Between October 2000 and February 2006, Dr. Pietrzyk served as a senior specialist of radio systems at Polska Telefonia Cyfrowa (PTC), a major mobile operator in Poland. He was responsible for evaluation of new radio technologies and functionalities before their introduction into the network, project management, and preparation of technology business cases. In 2003 he was suggested by his peer colleagues for the best employee award at PTC in the category "innovator," while later in 2005 in the category "team player." He contributed to the strategy definition for packet data network development and coordinated various projects including technology tests and trials.

Prior to PTC, Dr. Pietrzyk worked in academia in the Netherlands and in Poland, including participation in the Ubiquitous Communications Program at the Delft University of Technology, where he was a scientific researcher on the subproject Base Station Network and Signal Processing, working on multiple access schemes for OFDM systems. In 1999 he spent a month at the Interuniversity Micro-Electronics Center (IMEC) in Leuven, Belgium, developing a multiple access simulator for wireless systems. During his M.Sc. studies he was a student trainee working for fixed telecom operators in Poland and Greece.

Dr. Pietrzyk has served as a reviewer in the area of wireless access for numerous conferences and journals. He is a member of IEEE Communications and IEEE Computer Societies. His professional interests concentrate on radio systems with a high potential for offering novel services to a society. This includes systems such as broadcast, positioning, RFID and in particular broadband wireless access. His main focus is on broadband wireless access systems employing OFDM/OFDMA.

All comments or improvement suggestions regarding this book may be directed to the author at S.Pietrzyk@InnovativeSolutions.pl or S.Pietrzyk@ieee.org.

Index

Recent Titles in the Artech House Mobile Communications Series

John Walker, Series Editor

3G CDMA2000 Wireless System Engineering, Samuel C. Yang

3G Multimedia Network Services, Accounting, and User Profiles, Freddy Ghys, Marcel Mampaey, Michel Smouts, and Arto Vaaraniemi

802.11 WLANs and IP Networking: Security, QoS, and Mobility, Anand R. Prasad, Neeli R. Prasad

Advances in 3G Enhanced Technologies for Wireless Communications, Jiangzhou Wang and Tung-Sang Ng, editors

Advances in Mobile Information Systems, John Walker, editor

Advances in Mobile Radio Access Networks, Y. Jay Guo

Applied Satellite Navigation Using GPS, GALILEO, and Augmentation Systems, Ramjee Prasad and Marina Ruggieri

CDMA for Wireless Personal Communications, Ramjee Prasad

CDMA Mobile Radio Design, John B. Groe and Lawrence E. Larson

CDMA RF System Engineering, Samuel C. Yang

CDMA Systems Capacity Engineering, Kiseon Kim and Insoo Koo

CDMA Systems Engineering Handbook, Jhong S. Lee and Leonard E. Miller

Cell Planning for Wireless Communications, Manuel F. Cátedra and Jesús Pérez-Arriaga

Cellular Communications: Worldwide Market Development, Garry A. Garrard

Cellular Mobile Systems Engineering, Saleh Faruque

The Complete Wireless Communications Professional: A Guide for Engineers and Managers, William Webb

EDGE for Mobile Internet, Emmanuel Seurre, Patrick Savelli, and Pierre-Jean Pietri

Emerging Public Safety Wireless Communication Systems, Robert I. Desourdis, Jr., et al.

The Future of Wireless Communications, William Webb

GPRS for Mobile Internet, Emmanuel Seurre, Patrick Savelli, and Pierre-Jean Pietri

GPRS: Gateway to Third Generation Mobile Networks, Gunnar Heine and Holger Sagkob

GSM and Personal Communications Handbook, Siegmund M. Redl, Matthias K. Weber, and Malcolm W. Oliphant

GSM Networks: Protocols, Terminology, and Implementation, Gunnar Heine

GSM System Engineering, Asha Mehrotra

Handbook of Land-Mobile Radio System Coverage, Garry C. Hess

Handbook of Mobile Radio Networks, Sami Tabbane

High-Speed Wireless ATM and LANs, Benny Bing

Interference Analysis and Reduction for Wireless Systems, Peter Stavroulakis

Introduction to 3G Mobile Communications, Second Edition, Juha Korhonen

Introduction to Communication Systems Simulation, Maurice Schiff

Introduction to Digital Professional Mobile Radio, Hans-Peter A. Ketterling

Introduction to GPS: The Global Positioning System, Ahmed El-Rabbany

An Introduction to GSM, Siegmund M. Redl, Matthias K. Weber, and Malcolm W. Oliphant

Introduction to Mobile Communications Engineering, José M. Hernando and F. Pérez-Fontán

Introduction to Radio Propagation for Fixed and Mobile Communications, John Doble

Introduction to Wireless Local Loop, Second Edition: Broadband and Narrowband Systems, William Webb

IS-136 TDMA Technology, Economics, and Services, Lawrence Harte, Adrian Smith, and Charles A. Jacobs

Location Management and Routing in Mobile Wireless Networks, Amitava Mukherjee, Somprakash Bandyopadhyay, and Debashis Saha

Mobile Data Communications Systems, Peter Wong and David Britland

Mobile IP Technology for M-Business, Mark Norris

Mobile Satellite Communications, Shingo Ohmori, Hiromitsu Wakana, and Seiichiro Kawase

Mobile Telecommunications Standards: GSM, UMTS, TETRA, and ERMES, Rudi Bekkers

Mobile Telecommunications: Standards, Regulation, and Applications, Rudi Bekkers and Jan Smits

Multiantenna Digital Radio Transmission, Massimiliano "Max" Martone

Multiantenna Wireless Communicatio ns Systems, Sergio Barbarossa

Multipath Phenomena in Cellular Networks, Nathan Blaunstein and Jørgen Bach Andersen

Multiuser Detection in CDMA Mobile Terminals, Piero Castoldi

OFDMA for Broadband Wireless Access, Sławomir Pietrzyk

Personal Wireless Communication with DECT and PWT, John Phillips and Gerard Mac Namee

Practical Wireless Data Modem Design, Jonathon Y. C. Cheah

Prime Codes with Applications to CDMA Optical and Wireless Networks, Guu-Chang Yang and Wing C. Kwong

QoS in Integrated 3G Networks, Robert Lloyd-Evans

Radio Engineering for Wireless Communication and Sensor Applications, Antti V. Räisänen and Arto Lehto

Radio Propagation in Cellular Networks, Nathan Blaunstein

Radio Resource Management for Wireless Networks, Jens Zander and Seong-Lyun Kim

RDS: The Radio Data System, Dietmar Kopitz and Bev Marks

Resource Allocation in Hierarchical Cellular Systems, Lauro Ortigoza-Guerrero and A. Hamid Aghvami

RF and Baseband Techniques for Software-Defined Radio Peter B. Kenington

RF and Microwave Circuit Design for Wireless Communications, Lawrence E. Larson, editor

Sample Rate Conversion in Software Configurable Radios, Tim Hentschel

Signal Processing Applications in CDMA Communications, Hui Liu

Smart Antenna Engineering, Ahmed El Zooghby

Software Defined Radio for 3G, Paul Burns

Spread Spectrum CDMA Systems for Wireless Communications, Savo G. Glisic and Branka Vucetic

Third Generation Wireless Systems, Volume 1: Post-Shannon Signal Architectures, George M. Calhoun

Traffic Analysis and Design of Wireless IP Networks, Toni Janevski

Transmission Systems Design Handbook for Wireless Networks, Harvey Lehpamer

UMTS and Mobile Computing, Alexander Joseph Huber and Josef Franz Huber

Understanding Cellular Radio, William Webb

Understanding Digital PCS: The TDMA Standard, Cameron Kelly Coursey

For further information on these and other Artech House titles, including previously considered out-of-print books now available through our In-Print-Forever® (IPF®) program, contact:

Artech House
685 Canton Street
Norwood, MA 02062
Phone: 781-769-9750
Fax: 781-769-6334
e-mail: artech@artechhouse.com

Artech House
46 Gillingham Street
London SW1V 1AH UK
Phone: +44 (0)20 7596-8750
Fax: +44 (0)20 7630-0166
e-mail: artech-uk@artechhouse.com

Find us on the World Wide Web at: www.artechhouse.com